Manufacturing Optimization through Intelligent Techniques

T0225497

MANUFACTURING ENGINEERING AND MATERIALS PROCESSING
A Series of Reference Books and Textbooks

SERIES EDITOR

Geoffrey Boothroyd
Boothroyd Dewhurst, Inc.
Wakefield, Rhode Island

Manufacturing Optimization through Intelligent Techniques

R. Saravanan

CRC Press
Taylor & Francis Group
Boca Raton London New York

CRC Press is an imprint of the
Taylor & Francis Group, an **informa** business

First published 2006 by CRC Press
Taylor & Francis Group
6000 Broken Sound Parkway NW, Suite 300
Boca Raton, FL 33487-2742

Reissued 2018 by CRC Press

© 2006 by Taylor & Francis
CRC Press is an imprint of Taylor & Francis Group, an Informa business

No claim to original U.S. Government works

This book contains information obtained from authentic and highly regarded sources. Reasonable efforts have been made to publish reliable data and information, but the author and publisher cannot assume responsibility for the validity of all materials or the consequences of their use. The authors and publishers have attempted to trace the copyright holders of all material reproduced in this publication and apologize to copyright holders if permission to publish in this form has not been obtained. If any copyright material has not been acknowledged please write and let us know so we may rectify in any future reprint.

Except as permitted under U.S. Copyright Law, no part of this book may be reprinted, reproduced, transmitted, or utilized in any form by any electronic, mechanical, or other means, now known or hereafter invented, including photocopying, microfilming, and recording, or in any information storage or retrieval system, without written permission from the publishers.

For permission to photocopy or use material electronically from this work, please access www.copyright.com (http://www.copyright.com/) or contact the Copyright Clearance Center, Inc. (CCC), 222 Rosewood Drive, Danvers, MA 01923, 978-750-8400. CCC is a not-for-profit organiza-tion that provides licenses and registration for a variety of users. For organizations that have been granted a photocopy license by the CCC, a separate system of payment has been arranged.

Trademark Notice: Product or corporate names may be trademarks or registered trademarks, and are used only for identification and explanation without intent to infringe.

A Library of Congress record exists under LC control number: 2005052197

Publisher's Note
The publisher has gone to great lengths to ensure the quality of this reprint but points out that some imperfections in the original copies may be apparent.

Disclaimer
The publisher has made every effort to trace copyright holders and welcomes correspondence from those they have been unable to contact.

ISBN 13: 978-1-138-10609-3 (hbk)
ISBN 13: 978-0-203-71320-4 (ebk)

Visit the Taylor & Francis Web site at http://www.taylorandfrancis.com and the
CRC Press Web site at http://www.crcpress.com

Dedicated to
My Mother
Chandra Kantha

Acknowledgments

I wish to express my personal gratitude to everyone who has made this book a reality. At the outset, I wish to convey my heartfelt thanks to Dr. V. Shanmuganathan, director, MAM College of Engineering and Technology, Tiruchirapalli, who was the initiator for my endeavor.

I am ever thankful to my research guide, Dr. M. Sachithanandam, principal, KLN College of Information Technology, Madurai.

I am very much thankful to Dr. N. Jawahar, Thiagarajar College of Engineering, Madurai; Dr. P. Asokan; and Dr. G. Prabhakar, National Institute of Technology, Tiruchirapalli, for their contributions to this book and for their encouragement and support given to me for writing this book.

I express my sincere thanks to Professor K. Sivakumar, Professor M. Venkatesa Prabu, Professor R. Maheswari, S. Ramabalan, A. Hussain Lal, and R. David Arockiaraj, who were my colleagues at J.J. College of Engineering and Technology, Trichirapalli, for their support throughout this project.

I wish to thank my students, Mr. Ranjith, J.J. College of Engineering and Technology, Tiruchirapalli; and Mr. Karthikeyan and Mr. Muthiah of the National Institute of Technology, Tiruchirapalli, for their enthusiasm.

I am very much thankful to my superiors at Kumaraguru College of Technology, Coimbatore, Dr. K. Arumugam, correspondent, Dr. A. Selvaraj, joint correspondent, and Dr. K.K. Padmanabhan, principal, and also to all my colleagues for their encouragement.

I am indebted to and thank all my family members who have helped me to achieve this accomplishment through their enduring patience and support. I thank my father, R. Rajendran, my wife, K. Chitradevi, my daughters, S. Dharani and S. Sangavi, my brother, R. Murali, and my sisters, R. Saradha Devi and R. Vani.

Finally, I am grateful to John Corrigan, acquisitions editor for Marcel Dekker Publications, and also to Cindy Renee Carelli, acquisitions editor for the Taylor & Francis Group and Preethi Cholmondeley and Marsha Pronin, both project coordinators for the Taylor & Francis Group.

Abstract

Manufacturing optimization is the systematic and scientific approach for solving problems concerned with various manufacturing functions to make the best production decisions. The ultimate goal of all such decisions is either to minimize the effort required or to maximize the desired benefit. Because the effort required or the benefit desired in any practical manufacturing situation can be expressed as a function of certain decision variables, manufacturing optimization can be defined as the process of finding certain decision variables that give the maximum or minimum value of one or more objective functions subject to some resources or process constraints. Any manufacturing function can be formulated as a multivariable, nonlinear, constrained optimization problem.

Many conventional optimization techniques are used for solving production problems. But all these techniques are not robust and each technique is suitable for solving a particular type of manufacturing optimization problem. To overcome the difficulties with the conventional techniques, the following intelligent techniques are described: genetic algorithm (GA), simulated annealing algorithm (SAA), particle swarm optimization (PSO), tabu search (TS), and ant colony optimization (ACO). These modern techniques are described here for solving the following manufacturing optimization problems: design of machine elements, machining tolerance allocation, selection of operating parameters for CNC machine tools, integrated product development, production scheduling, part family formation and machine grouping for cellular manufacturing and flexible manufacturing systems, robot trajectory planning, and intelligent manufacturing.

After reading this book, the reader will be able to understand the different types of manufacturing optimization problems and the conventional and intelligent techniques suitable for solving mentioned above. By understanding the concepts and the different types of optimization techniques, the reader will be able to develop and implement suitable optimization procedures and algorithms for a wide variety of problem domains in the area of design and manufacturing.

Author

Dr. R. Saravanan earned a B.E. in mechanical and production engineering in 1985 and an M.E in production engineering in 1992 from Annamalai University, Chidambaram, India; and in 2001, a Ph.D. in computer-aided manufacturing from the National Institute of Technology (REC), Tiruchirapalli, India.

Dr. Saravanan has 18 years of experience in industry, teaching and research, including two years as a production superintendent at SRP Tools Ltd., Chennai; 2 years as a scientific assistant at the Indian Institute of Science, Bangalore; and 14 years as a teacher and researcher at engineering colleges. Currently, Dr. Saravanan is a professor and head of the Department of Mechatronics Engineering at Kumaragura College of Technology, Coimbatore, Tamilnadu, India.

He has presented 30 research papers at national and international conferences and has published 14 papers in international journals and five papers in national journals. He is also a research guide at Anna University, Bharathidasan University, Sri Chandrasekarendra Saraswathi Mahavidyalaya-Kancheepuram, and SASTRA-Thanjavur (Deemed Universities). Dr. Saravanan currently is guiding 8 research scholars for Ph.D. degrees.

The Technology Information Forecasting and Assessment Council (TIFAC), New Delhi, recognizes Dr. Saravanan as an expert member. He has also received the Gold Medal–Best Paper award (2001) from the Institution of Engineers in India and the Bharath Vikas award (2002) from the Integrated Council of Socio-Economic Progress, New Delhi. In 2003, Dr. Saravanan was included in *Marquis Who's Who*, U.S.A.

Contents

1 Manufacturing Optimization through Intelligent Techniques

Optimization is the science of getting best results subject to various resource constraints. When optimization is applied to different manufacturing functions, we call this subject manufacturing optimization. There is a lot of scope for optimizing various manufacturing functions such as design, planning, operation, quality control, maintenance, and so forth. Nowadays, professionals from academic institutions and industries have started realizing the importance of this new manufacturing optimization, in order to improve performance. In the past, due to lack of computational facilities and optimization techniques, few attempts were made to formulate the various manufacturing activities as optimization problems and to develop the procedures for the same. Today, many computing facilities and numerous non-conventional techniques—particularly intelligent techniques—are available for constructing the mathematical model and to develop the procedure and software. It is proposed to formulate the various manufacturing activities as optimization problems and to use different intelligent techniques for solving the same.

Many conventional optimization techniques are used for solving different manufacturing optimization problems. Not all these techniques are robust and each technique is suitable for solving a particular type of manufacturing optimization problem. Most traditional optimization methods are appropriate for well-behaved, unimodal, simple objective functions; when applied to multimodal problems or problems where gradient information is not available, most of these methods either cannot be used or are not very efficient. This result suggests that better methods are required to solve complex, multimodal, discrete, or discontinuous problems. In general, we are interested in robust search techniques that can easily apply to a wide variety of problems. Most traditional methods are not robust because each of them is specialized to solve a particular class of problem, which is why so many different types of optimization methods exist. For different problems, different algorithms must be applied.

However, this discussion does not mean that these traditional algorithms are useless; in fact, they have been used extensively in many engineering optimization problems. If the solutions obtained by some traditional methods are satisfactory, no problem exists. If the solutions obtained are not satisfactory or some known methods cannot be applied, the user either must learn and use some other conventional optimization technique suitable to solve that problem (by no means an

1

easy matter) or the user must know some robust search algorithm that can be applied to a wide variety of problems without much difficulty.

To overcome the difficulties with the conventional techniques, the following intelligent techniques are described: genetic algorithm (GA), simulated annealing algorithm (SAA), particle swarm optimization (PSO), tabu search (TS) and ant colony optimization (ACO). These modern techniques are described for solving the following manufacturing optimization problems: design of mechanical elements, machining tolerance allocation, selection of operating parameters for CNC machine tools, integrated product development; production scheduling, part family formation and machine grouping for cellular manufacturing and flexible manufacturing systems, robot trajectory planning, and intelligent manufacturing.

Chapter 2 describes different conventional techniques suitable for solving various manufacturing optimization problems. Chapter 3 describes different nonconventional techniques. The remaining chapters describe various manufacturing optimization problems with suitable examples and discuss in detail applications of various conventional and nonconventional techniques.

2 Conventional Optimization Techniques for Manufacturing Applications

2.1 BRIEF OVERVIEW OF TRADITIONAL OPTIMIZATION TECHNIQUES

Most traditional optimization methods used for manufacturing applications can be divided into two broad classes: direct search methods requiring only the objective function values and gradient search methods requiring gradient information either exactly or numerically. One common characteristic of most of these methods is that they all work on a point-by-point basis. An algorithm starts with an initial point (usually supplied by the user) and depending on the transition rule used in the algorithm, a new point is determined. Essentially, algorithms vary according to the transition rule used to update a point.

Among the direct search methods, pattern search methods and conjugate direction methods have been used extensively. In pattern search methods, at every iteration a search direction is created according to a combination of a local exploratory search and a pattern search regulated by some heuristic rules. This method is often terminated prematurely and degenerates into a sequence of exploratory moves even though a number of fixes exist for some of these problems. Whether added complications are worth the trouble in such heuristic methods is questionable.

In conjugate direction methods, a set of conjugate search directions are generated using the history of a few previous iterations. Even though this method is very popular, the common problem with this method is that often the search directions become independent and occasional restarts are necessary. Moreover, this algorithm has a convergence proof for well-behaved, unimodal functions. In case of multimodal problems or skewed objective functions, the obtained solution depends largely on the initial point.

Box's direct search method differs from these methods in that the algorithm works with a number of points instead of a single point. The algorithm resumes with an evenly distributed set of points. At every iteration, a new set of points is created around the best point of the previous iteration. Since no information about the rejected points is used in choosing new points in subsequent iterations, the method is slow and not very efficient; but if widespread points are used, the algorithm may eventually find the global solution but the time to obtain the global solution may be too large to make the search useful in real-world problems.

The simplex search method uses a simplex of points to create a new simplex according to some rules that depend on the objective function values at all points of the simplex. The essential idea is to generate a search direction using the simplex. Since the whole search space cannot be spanned with one search direction, the simplex search is blind and generally cannot find the global solution.

In general, direct search methods are expensive and seem to work on simple unimodal functions. Gradient-based methods, on the other hand, require the knowledge or gradients of functions and constraints. Again, at least two difficulties exist with these methods. Most of these algorithms are not guaranteed to find the global optimal solutions because these algorithms usually terminate when the gradient of the objective function is very close to zero, which may happen both in case of local and global solutions. The other difficulty is the calculation of the gradients themselves. In most real-world engineering design problems, the objective function is not explicitly known. Some simulation (running a finite element package, for example) is required to evaluate the objective function or constraints. Thus, the exact computation of the gradient may not be possible in some problems. Even though gradients can be evaluated numerically, they are not exact. Some methods require the computation of a Hessian matrix, the numerical computation of which is expensive and not accurate. Moreover, if some of the design variables are integers, numerical gradient computation becomes difficult.

Some random search techniques are also used extensively in problems where no knowledge about the problem is known, where the search space is large or where none of the traditional methods have worked. These methods are also used to find a feasible starting point, especially if the number of constraints is large. In these methods, a number of points in a predefined range is created at random and the iteration proceeds by creating a new set of points in a narrower region around the best point found among the current set of points. As iteration progresses, the search region narrows. Since the points are created at random, the reliability of these methods depends on the number of points considered in each region and the number of steps used. If more points per step and more steps are used, the accuracy will increase but the computational complexity will be large. Thus, these methods are usually expensive and used as a last resort.

In this chapter, only direct search techniques found in the literature for solving various manufacturing optimization problems are described.

2.2 SINGLE VARIABLE TECHNIQUES SUITABLE FOR SOLVING VARIOUS MANUFACTURING OPTIMIZATION PROBLEMS (DIRECT SEARCH METHODS)

In most optimization problems, the solution is known to exist within the variable bounds. In some cases, this range is not known and hence the search must be made with no restrictions on the values of the variables.

2.2.1 Unrestricted Search

The search must be made with no restriction on the values of the variables. The simplest approach for such a problem is to use a fixed step size and move in a favorable direction from an initial guess point. The step size used must be small in relation to the final accuracy desired. Although this method is very simple to implement, it is not efficient in many cases.

2.2.2 Search with Fixed Step Size

2.2.2.1 Steps

Start with an initial guess point, say, x_1.
Find $f_1 = f(x_1)$.
Assuming a step size s, find $x_2 = x_1 + s$.
Find $f_2 = f(x_2)$.
If $f_2 < f_1$ and if the problem is one of minimization, the assumption of unimodality indicates that the desired minimum cannot lie between x_1 and x_2. Hence, the search can be continued further along using points $x_3, x_4\ldots$ using the unimodality assumption while testing each pair of experiments. This procedure is continued until a point, $x_i = x_{i1} + s$, shows an increase in the function value.
The search is terminated at x_i and either x_{i1} or x_i can be taken as the optimum point.
Originally, if $f_2 > f_1$, the search should be carried in the reverse direction at points $x_2, x_3\ldots$ where $x_j = x_1$ $(j1)s$.
If $f_2 = f_1$, the desired minimum lies between x_1 and x_2 and the minimum point can be taken as either x_1 or x_2.
If both f_2 and f_2 are greater than f_1, this condition implies the desired minimum will lie in the double interval $x_2 < x < x_2$.

The major limitation is the unrestricted nature of the region in which the minimum can lie.

2.2.3 Search with Accelerated Step Size

The major limitation of fixed step size is the unrestricted nature of the region in which the minimum can lie. This unrestricted region involves a large amount of computational time that can be minimized by increasing the step size gradually until the minimum point is bracketed.

2.2.4 Exhaustive Search Method

The exhaustive search method can be used to solve problems where the interval in which the optimum is known to lie is finite. This method consists of evaluating the objective function at a predetermined number of equally spaced points in

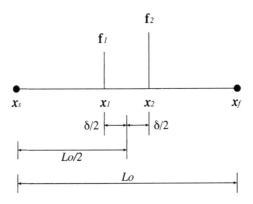

FIGURE 2.1 Dichotomous search.

the given interval and reducing the interval of uncertainty using the assumption of unimodality.

2.2.5 DICHOTOMOUS SEARCH

The dichotomous search is a sequential search method shown in Figure 2.1. Results of any experiment influence the location of the subsequent experiment. Two experiments are placed as close as possible at the center of the interval of uncertainty. Based on the relative values of the objective function of the two points, almost half the interval of uncertainty is eliminated. A pair of experiments are conducted at the center of the current interval of uncertainty; the next pair, at the center of the remaining interval of uncertainty,

$$x_1 = L_o/2 - \delta/2$$

$$x_2 = L_o/2 + \delta/2$$

where L_o is the initial interval and δ is the minimum value for conducting the experiment.

2.2.6 FIBONACCI SEARCH

The Fibonacci search is a very efficient sequential technique. It is based on the use of the Fibonacci number series named after a 13th-century mathematician.

A Fibonacci series is given by $F_n = F_{n2} + F_{n-1}$, where $F_o = 1$ and $F_1 = 1$. The series is

$$n = 0, 1, 2, 3, 4, 5, 6, 7, 8, 9\dots$$

$$F_n = 1, 1, 2, 3, 5, 8, 13, 21, 34, 55\dots$$

Note that the nth Fibonacci number is the sum of the two preceding numbers. The location of the first two experiments is determined by number of experiments, n.

2.2.7 Disadvantages

A Fibonacci search requires an advance decision on the number of experiments before any information about the behavior of the function near the maximum is obtained.

2.2.8 Golden Search Method

The golden section method does not require an advance decision on the number of trials. This technique is based on the fact that the ratio of the two successive Fibonacci numbers for all values of n is greater than 8. This ratio was discovered by Euclid and is called the Golden Mean. The procedure is same as the Fibonacci method except that the location of the first two trials are located at 0.618 L from the end of this range to eliminate the interval, where L is the interval of uncertainty. Based on the value of y_1 and y_2 and in the new reduced interval, perform additional experiments at $\pm$ 0.618 L 2.

2.3 MULTIVARIABLE TECHNIQUES SUITABLE FOR SOLVING VARIOUS MANUFACTURING OPTIMIZATION PROBLEMS (DIRECT SEARCH METHODS)

In this section, a number of minimization algorithms are presented that use the function value only. If the gradient information is valuable, a gradient-based method can be more efficient. Unfortunately, many real-world optimization problems require the use of computationally expensive simulation packages to calculate the objective function, thereby making it difficult to compute the derivative of the objective function. In these problems, direct search techniques may be useful.

In a single-variable function optimization, only two search directions exist in which a point can be modified — either in the positive x-direction or the negative x-direction. The extent of increment or decrement in each direction depends on the current point and the objective function. In multi-objective function optimization, each variable can be modified either in the positive or the negative direction, thereby totaling 2^N different ways. Moreover, an algorithm having searches along each variable one at a time can only successfully solve linearly separable functions. These algorithms (called "one-variable-at-a-time methods") cannot usually solve functions having nonlinear interactions among design variables. Ideally, algorithms are required that either completely eliminate the concept of search direction and manipulate a set of points to create a better set of points or use complex search directions to effectively decouple the non-linearity of the function. In the following section, four algorithms applied to different manufacturing optimization problems are described.

2.3.1 Evolutionary Optimization Method

Evolutionary optimization is a simple optimization technique developed by G.E.P. Box in 1957. The algorithm requires $(2^N + 1)$ points, of which 2^N are corner points of an N-dimensional hypercube centered on the other point. All $(2^N + 1)$ function values are compared and the best point is identified. In the next iteration, another hypercube is found around this best point. If at any iteration an improved point is not found, the size of the hypercube is reduced. This process continues until the hypercube becomes very small.

2.3.1.1 Algorithm

Step 1: Choose an initial point $X^{(0)}$ and size reduction parameters Δ for all design variables, $i = 1, 2, 3...N$. Choose a termination parameter, ε. Set $\bar{x} = X^{(0)}$.

Step 2: If $\|\Delta\| < \varepsilon$, terminate; else create 2^N points by adding and subtracting $\Delta i/2$ from each variable at the point $\bar{x}$.

Step 3: Compute function values at all $(2^N + 1)$ points. Find the point having the minimum function value. Designate the minimum point to be X.

Step 4: If $\bar{x} = X^{(0)}$, reduce size parameters $\Delta i = \Delta i/2$ and go to Step 2; else set $X^{(0)} = \bar{x}$ and go to Step 2.

In the above algorithm, $X^{(0)}$ is always set as the current best point. Thus, at the end of simulation, $X^{(0)}$ becomes the obtained optimum point. It is evident from the algorithm that at most 2^N functions are evaluated at each iteration. Thus, the required number of function evaluations increases exponentially with N. The algorithm, however, is simple to implement and has had success in solving many industrial optimization problems.

2.3.2 Nelder–Mead Simplex Method

In the simplex search method, the number of points in the initial simplex is three for 2 variables only. Even though some guidelines to choose the initial simplex are suggested, the points chosen for the initial simplex should not form a zero-volume N-dimensional hypercube. Thus, in a function with two variables, the chosen three points in the simplex should not lie along a line. Similarly, in a function with three variables, four points in the initial simplex should not lie on a plane.

At each iteration, the worst point in the simplex is found first. A new simplex is then formed from the old simplex by some fixed rules that steer the search away from the worst point in the simplex. The extent of steering depends on the relative function values of the simplexes. Four different situations can arise depending on the function values. The situations are depicted in Figure 2.2.

At first, the initial simplex is formed and is designated according to its performance. The best point in the current simplex is designated as X_l, the worst point is X_h and the next best point is X_g. The centroid (X_c) of all but the worst point is determined,

$$X_c = (X_l + X_g)/2.$$

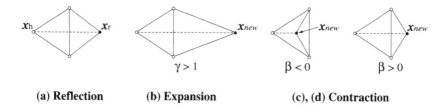

	$\gamma > 1$	$\beta < 0$ $\beta > 0$
(a) Reflection	**(b) Expansion**	**(c), (d) Contraction**

FIGURE 2.2 An illustration of the simplex search method. First, a reflection is performed (a). Depending on function values, an expansion (b) and two different contractions, (c) and (d), are possible.

Thereafter, the worst point in the simplex is reflected about the centroid and a new point X_r is found,

$$X_r = 2X_o - X_h.$$

The reflection operation is depicted in Figure 2.2a. If the function value at this point is better than the best point in the simplex, the reflection is considered to have taken the simplex to a good region in the search space. Thus, an expansion along the direction from the centroid to the reflected point is performed (Figure 2.2b),

$$X_{new} = (1 + \gamma) X_o - \gamma X_h,$$

where γ is the expansion coefficient.

On the other hand, if the function value at the reflected point is worse than the worst point in the simplex, the reflection is considered to have taken the simplex to a bad region in the search space. Thus, a contraction in the direction from the centroid to the reflected point is made (Figure 2.2c),

$$X_{new} = (1 + \beta) X_o - \beta X_h,$$

where β is the contraction coefficient.

The amount of contraction is controlled by the factor β. Finally, if the function value at the reflected point is better than the worst and worse than the next-to-worst point in the simplex, a contraction is made with a positive β value (Figure 2.2d). The default scenario is the reflected point itself. The obtained new point replaces the worst point in the simplex and the algorithm continues with the new simplex. This algorithm was originally proposed by Nelder and Mead (1965).

2.3.2.1 Algorithm

Step 1: Choose $\gamma > 1$, $\in (0, 1)$, and a termination parameter, ϵ. Create an initial simplex.

Step 2: Find X_h (the worst point), X_l (the best point) and X_g (next-to-worst point). Calculate

$X_c = (1/N) \sum X_i.$

Step 3: Calculate the reflected point

$X_r = 2 X_c - X_h$, set $X_{new} = X_r$

If f (X_r) < f (X_l), set $X_{new} = (1 + \gamma) X_c - \gamma X_h$ *(expansion)*

   Else if. f (X_r) > f $(X_h,)$, *set $X_{new} = (1 - \beta) X_o + \beta X_h$ (contraction)*

   Else if . f (X_g) < f (X_r), *set $X_{new} = (1 + \beta) X_o - \beta X_h$ (contraction)*

   Calculate f (X_{new}) and replace X_h by X_{new}

Step 4: *if* $\left\{ \sum\limits_{i=1}^{N+1} \dfrac{(f(x_i) - f(x_c))}{N+1} \right\}^{1/2} \leq \epsilon, Ter \min ate;$

else go to Step 2.

Any other termination criteria can also be used. The performance of the above algorithm depends on the values of β and γ. If a large value of γ or $1/\beta$ is used, the approach to the optimum point may be faster but the convergence to the optimum point may be difficult. Smaller values of γ or $1/\beta$ may require more function evaluations to converge near the optimum point. The recommended values for parameters are $\gamma \sim 2.0$ and $|\beta| \sim 0.5$.

2.3.3 COMPLEX METHOD

The complex search method is similar to the simplex method except the constraints are not handled in the former method. This method was developed by M.J. Box in 1965. The algorithm begins with a number of feasible points created at random. If a point is found to be infeasible, a new point is created using the previously generated feasible points. Usually, the infeasible point is pushed towards the centroid of the previously found feasible points. Once a set of feasible points is found, the worst point is reflected about the centroid of the rest of the points to find a new point. Depending on the feasibility and function value of the new point, the point is further modified or accepted. If the new point falls outside the variable boundaries, the point is modified to fall on the violated boundary. If the new point is feasible, the point is retracted towards the feasible points. The worst point in the simplex is replaced by this new feasible point and the algorithm continues for the next interaction.

2.3.4 HOOKE–JEEVES PATTERN SEARCH METHOD

The pattern search method works by iteratively creating a set of search directions. The created search directions should be such that they completely span the search space. In other words, they should be such that starting from any point in the search space, any other point in the search space can be reached by traversing along these search directions only. In an N-dimensional problem, this requires at least N linearly independent search directions. For example, in a two-variable function, at least two search directions are required to go from any one point to any other point. Among many possible combinations of N search directions, some combinations may be able to reach the destination faster (with fewer iterations) and some may require more iterations.

In the Hooke–Jeeves method, a combination of exploratory moves and heuristic pattern moves is made iteratively. An exploratory move is performed systematically in the vicinity of the current point to find the best point around the current point. Thereafter, two such points are used to make a pattern move. Each of these moves is described in the following sections.

2.3.4.1 Exploratory Move

Assume that the current solution (the base point) is denoted by X_c. Assume also that the variable X_c is perturbed by i. Set $i = 1$ and $x = X_c$.

Step 1: Calculate $f = f(x)$, $f^+ = f(x_i + \Delta i)$, and $f^- = f(x_i - \Delta i)$.
Step 2: Find $f_{min} = \min(f, f^+, f^-)$. Set x to correspond to f_{min}.
Step 3: Is $i = N$? If no, set $i = i + 1$ and go to Step 1; else x is the result and go to Step 4.
Step 4: If $x \neq X_c$, success; else failure.

In the exploratory move, the current point is perturbed in positive and negative directions along each variable one at a time and the best point is recorded. The current point is changed to the best point at the end of each variable perturbation. If the point found at the end of all variable perturbations is different than the original point, the exploratory move is a success; otherwise, the exploratory move is a failure. In any case, the best point is considered to be the outcome of the exploratory move.

2.3.4.2 Pattern Move

A new point is found by jumping from the current best point X_c along a direction connecting the previous best point $X_{(k-1)}$ and the current base point $X_{(k)}$ as follows:

$$X_{p(k+1)} = X_{(k)} + [X_{(k)} - X_{(k-1)}].$$

The Hooke–Jeeves method consists of an iterative application of an exploratory move in the locality of the current point as a subsequent jump using the pattern move. If the pattern move does not take the solution to a better region, the pattern move is not accepted and the extent of the exploratory search is reduced.

2.3.4.3 Algorithm

Step 1: Choose a starting point $X_{(0)}$, variable increments i ($i = 1, 2, 3...N$), a step reduction factor $\alpha > 1$, and a termination parameter, ε. Set $k = 0$.
Step 2: Perform an exploratory move with $X_{(k)}$ as the base point. Say x is the outcome of the exploratory move. If the exploratory move is a success, set $X_{(k+1)} = x$ and go to Step 4.

Step 3: Is $|\Delta| < \varepsilon$? if yes, terminate; else set $i/2$ for $i = 1, 2, 3...N$ and go to Step 2.

Step 4: Set $k = k + 1$ and perform the pattern move:

$$X_{(k+1)} = X_{(k)} + [X_{(k)} - X_{(k-1)}]$$

Step 5: Perform another exploratory move using $X_{p(k+1)}$ as the base point. Let the result be $X_{(k+1)}$.

Step 6: Is $f[X_{(k+1)}] < X_{(k)}$? If yes, go to Step 4; else go to Step 3.

The search strategy is simple and straightforward. The algorithm requires less storage for variables; only two points $X_{(k)}$ and $X_{(k-1)}$ need to be stored at any iteration. The numerical calculations involved in the process are also simple. But because the search depends largely on the moves along the coordinate directions (x_1, x_2, and so on) during the exploratory move, the algorithm may prematurely converge to a wrong solution, especially in the case of functions with highly nonlinear interactions among variables. The algorithm can also get stuck in the loop of generating exploratory moves either between Steps 5 and 6 or between Steps 2 and 3. Another feature of this algorithm is that it terminates only by searching exhaustively the vicinity of the converged point. This behavior requires a large number of function evaluations for convergence to a solution with a reasonable degree of accuracy. The convergence to the optimum point depends on the parameter α; a value $\alpha = 2$ is recommended.

2.4 DYNAMIC PROGRAMMING TECHNIQUE

In most practical problems, decisions must be made sequentially at different points in time, at different points in space, and at different levels for a component, a subsystem or a system. The problems in which the decisions are to be made sequentially are called sequential decision problems. Because these decisions are to be made at a number of stages, they are also referred to as multistage decision problems. Dynamic programming is a mathematical technique well suited for the optimization of multistage decision problems. This technique was developed by Richard Bellman in the early 1950s.

The dynamic programming technique, when applicable, represents or decomposes a multistage decision problem as a sequence of single stage decision problems. Thus, an N-variable problem is represented as a sequence of N single variable problems that are solved successively. In most of the cases, these N subproblems are easier to solve than the original problem. The decomposition to N subproblems is done in such a manner that the optimal solution of the original N-variable problem can be obtained from the optimal solutions of the N one-dimensional problems. The particular optimization technique used for the optimization of the N-single variable problems is irrelevant. It may range from a simple enumeration process to a differential calculus or a nonlinear programming technique.

Multistage decision problems can also be solved by the direct application of the classical optimization techniques. However, this procedure requires the number of variables to be small, the functions involved to be continuous and continuously differentiable and the optimum points to not lie at the boundary points.

Further, the problem must be relatively simple so the set of resultant equations can be solved either analytically or numerically. The nonlinear programming techniques can be used to solve slightly more complicated multistage decision problems but their application requires the variables to be continuous and a prior knowledge about the region of the global minimum or maximum. In all these cases, the introduction of stochastic variability makes the problem extremely complex and renders the problem unsolvable except by using some sort of an approximation, such as chance constrained optimization.

Dynamic programming, on the other hand, can deal with discrete variables, and with nonconvex, noncontinuous, and nondifferentiable functions. In general, it can also take into account the stochastic variability by a simple modification of the deterministic procedure. The dynamic programming technique suffers from a major drawback known as the "curse of dimensionality." However, in spite of this disadvantage, it is very suitable for the solution of a wide range of complex problems in several areas of decision making.

2.4.1 REPRESENTATION OF MULTISTAGE DECISION PROCESS

Before considering the representation of a multistage decision process, consider a single-stage decision process (a component of the multistage process) represented as a rectangular block in Figure 2.3. Any decision process is characterized by certain input parameters, S (or data), certain decision variables (X) and certain output parameters (T) representing the outcome obtained as a result of making

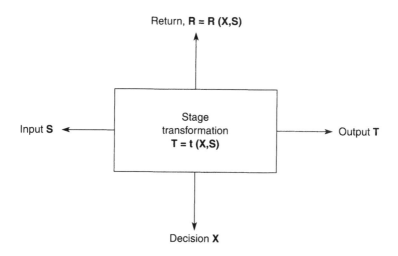

FIGURE 2.3 Single-stage decision problem.

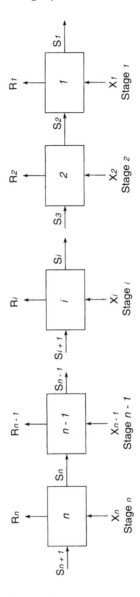

FIGURE 2.4 Multistage decision problem.

the decision. The input parameters are called input state variables and the output parameters are called output state variables. Finally, a return or objective function R exists that measures the effectiveness of the decisions that are made and the outputs that result from these decisions. For any physical system, that function is represented as a single stage decision process shown in Figure 2.3. Any serial multistage decision process can now be represented schematically, as shown in Figure 2.4. Because of some convenience, the stages n, $(n - 1)...3$, 2, 1 are labelled in decreasing order.

The objective of a multistage decision problem is to find $X_1, X_2...X_n$ so as to optimize some function of the individual stage returns, $f(R_1, R_s...R_a)$ and satisfy the equations. The nature of the n stage return function, f, determines whether a given multistage problem can be solved by dynamic programming. Because the method works as a decomposition technique, it requires the separability and monotonicity of the objective function.

The computational procedure is not described in this book. For more details of the dynamic programming and other techniques given in this chapter, readers can refer to the books and journals given in the references.

REFERENCES

Agapiou, J.S., The optimization of machining operations based on a combined criterion, part 1: the use of combined objectives in single pass operations, Transactions of *ASME: Journal of Engineering for Industry,* 114, 500–507, 1992.

Agapiou, J.S, The optimization of machining operations based on combined criterion, part 2: multipass operations, *ASME Journal of Engineering for industry,* 114, 508–513, 1992.

Arora, J., *Introduction to Optimum Design,* McGraw-Hill International, Singapore, 1989.

Box, G.E.P., Evolutionary operation: A method of increasing industrial productivity, *Applied Statistics,* 6, 81–101, 1957.

Box, M.J., A new method of constrained optimization and comparison with other methods, *Computer Journal,* Vol. 7, 42–52, 1965.

Deb, K., *Optimization for Engineering Design — Algorithms and Examples,* Prentice Hall of India (Pvt.) Ltd., New Delhi, 1998.

Nelden, J.A. and Mead, R., A simplex method for function minimization, *Computer Journal,* 7, 308–313, 1965.

Proceedings of two-week short term training program, "Modeling and optimization of manufacturing systems using conventional and non-conventional techniques," J.J. College of Engineering and Technology, Trichirapalli, India, 2003.

Rao, S.S., *Engineering Optimization: Theory and Practice,* Wiley Eastern, New Delhi, 1984

Saravanan, R. et al., Comparative analysis of conventional and non-conventional optimisation techniques for CNC turning process, *International Journal of Advanced Manufacturing Technology,* 17(7), 471–476, 2001.

3 Intelligent Optimization Techniques for Manufacturing Optimization Problems

In this chapter, the following intelligent (nonconventional) optimization techniques, reportedly successful in solving a wide variety of search and optimization problems in science, engineering and commerce, are described:

Genetic algorithm (GA)
Simulated annealing algorithm (SAA)
Particle swarm optimization (PSO)
Tabu search (TS)
Ant colony optimization (ACO)

Implementation of these techniques is described in Chapter 4 through Chapter 9 for different manufacturing optimization problems.

3.1 GENETIC ALGORITHMS (GA)

Genetic algorithms (GA) are adaptive search and optimization algorithms that mimic the principles of natural genetics. GAs are very different from traditional search and optimization methods used in different manufacturing problems. Because of their simplicity, ease of operation, minimal requirements, and global perspective, GAs have been successfully used in a wide variety of problem domains. GAs were developed by John Holland of the University of Michigan in 1965.

3.1.1 WORKING PRINCIPLE OF GA

Genetic algorithms are search and optimization procedures motivated by the principles of natural genetics and natural selection. Some fundamental ideas of genetics are borrowed and used artificially to construct search algorithms that are robust and require minimal problem information. The working principles of GAs are very different from that of most of traditional optimization techniques and are given in Figure 3.1. Here, the working principles of GAs are first described. An unconstrained, single-variable optimization problem is given below:

Begin

Initialize population;

Evaluate population;

Repeat

{

Reproduction;

Crossover;

Mutation;

}

Until (termination criteria);

End.

FIGURE 3.1 A pseudo-code for a simple genetic algorithm.

Maximize $f(x)$

Variable bound $X_{min} \geq X \geq X_{max}$

To use GA to solve the above problem, the variable X is typically coded in some string structures. Binary coded strings are mostly used. The length of the string is usually determined according to the accuracy of the solution desired. For example, if five-bit binary strings are used to code the variable X, the string [0 0 0 0 0] is coded to the value X_{min}, the string [1 1 1 1 1] is coded to the value X_{max} and any other string is uniquely coded to a value in the range (X_{min}, X_{max}). With five bits in a string, only 2 or 32 different strings are possible because each bit position can take a value of 0 or 1. In practice, strings of sizes ranging from a hundred to a few hundred places are common; recently, a coding with a string size equal to 16,384 has been used. Thus, with five-bit strings used to code the variable X, the accuracy between two consecutive strings is only $(X_{max} - X_{min})/31$. If more accuracy is desired, longer strings may be used. As the string length increases, the minimum possible accuracy in the solution increases exponentially. With a known coding, any string can be decoded to an X value that then can be used to find the objective function value. A string's objective function value, $f(x)$, is known as the string's fitness.

GAs begin with a population of string structures created at random. Thereafter, each string in the population is evaluated. The population is then operated by three main operators — reproduction, crossover, and mutation — to hopefully create a better population. The population is further evaluated and tested for termination. If the termination criteria are not met, the population is again operated by the above three operators and evaluated. This procedure is continued until the termination criteria are met. One cycle of these operators and the evaluation procedure is known as a generation in GA terminology.

Reproductions are usually the first operator applied on a population. Reproduction selects good strings in a population and forms a mating pool. A number of reproduction operators exist in GA literature but the essential idea is that above-average strings are picked from the current population and duplicates of them are inserted in the mating pool. The commonly used reproduction operator is the proportionate selection operator, where a string in the current population is selected with a probability proportional to the string's fitness. Thus, the ith string in the population is selected with a probability proportional to f. Since the population size is usually kept fixed in a simple GA, the cumulative probability for all strings in the population must be 1. Therefore, the probability for selecting the ith string is f/f_{avg}.

One way to achieve this proportionate selection is to use a roulette wheel with the circumference marked for each string proportionate to the string's fitness. Because the circumference of the wheel is marked according to a string's fitness, this roulette wheel mechanism is expected to make copies of the ith string, where f_{avg} is the average fitness of the population. Even though this version of roulette wheel selection is somewhat noisy, it is widely used. Other more stable versions of this roulette wheel selection also exist.

The crossover operator is applied next to the strings of the mating pool. A number of crossover operators exist in GA literature but in almost all crossover operators, two strings are picked from the mating pool at random and some portion of the stings are exchanged between the strings. In a single-point crossover operator, this operation is performed by randomly choosing a crossing site between the strings and exchanging all bits on the right side of the crossing site as shown below:

Before crossover:
0 0 0   0 0 0
1 1 1   1 1 1
After crossover:
0 0 0   1 1 1
1 1 1   0 0 0

Good substrings from either parent string can be combined to form a better child string if an appropriate site is chosen. Because the knowledge of an appropriate site is usually not known (a random site), the child strings produced may or may not have a combination of good substrings from the parent strings, depending on whether or not the crossing site falls in the appropriate place. This aspect is not worth worrying about very much because if good strings are created by crossover, more copies of them will exist in the next mating pool generated by the reproduction operator. If good strings are not created by crossover, they will not survive beyond the next generation because reproduction will not select those strings for the next mating pool.

In a two-point crossover operator, two random sites are chosen and the contents bracketed by these sites are exchanged between two parents. This idea can be extended to create a multipoint crossover operator and the extreme of this extension is known as a uniform crossover operator. In a uniform crossover for binary strings, each bit from either parent is selected with a probability of 0.5. One other aspect is that the search must be performed in a way that the information stored in the parent strings are maximally preserved because these parent strings are instances of good selection using the reproduction operator.

In the single-point crossover operator, the search is not extensive, but the maximum information is preserved from parent to child. In the uniform crossover, the search is very extensive, but minimum information is preserved between parent and child strings. Even though some studies to find an optimal crossover operator exist, considerable doubts remain about whether those results can be generalized for all problems. Before any results from theoretical studies are obtained, the choice of crossover operator is still a matter of personal preference. However, to preserve some good qualities in the mating pool, not all strings in the population are used in crossover. If a crossover probability of $P_c = 0$ is used, 100% of the population is simply copied to the new population.

The crossover operator is mainly responsible for the search aspect of genetic algorithms, even though the mutation operator is also sparingly used for this purpose. The mutation operator changes 1 to 0 and vice versa with a small mutation probability, P_m. The need for mutation is to keep diversity in the population. For example, if at a particular position along the string length all strings in the population have a value 0 and a 1 is needed in that position to obtain the optimum, neither the reproduction nor crossover operator described above is able to create a 1 in that position. The inclusion of mutation introduces some probability of turning that 0 into a 1. Furthermore, for local improvement of a solution, mutation can be useful.

These three operators are simple and straightforward. The reproduction operator selects good strings and the crossover operator recombines good substrings from two good strings together to hopefully form a better substring. The mutation operator alters a string locally to hopefully create a better string. Even though none of these claims are guaranteed or tested while creating a string, if bad strings are created they are expected to be eliminated by the reproduction operator in the next generation and if good strings are created, they will be emphasized.

3.1.1.1 Two-Point Crossover

In a two-point crossover operator, two random sites are chosen and the contents bracketed by these sites are exchanged between two mated parents. If the cross-site 1 is three and cross-site 2 is six, the strings between three and six are exchanged, as shown in Figure 3.2.

3.1.1.2 Multipoint Crossover

In a multipoint crossover, again two cases exist. One is an even number of cross-sites and the second is an odd number of cross-sites. In the case of even numbered cross-sites, the string is treated as a ring with no beginning or end. The cross-sites are uniformly selected around the circle at random. Now the information between alternate pairs of sites is interchanged, as shown in Figure 3.3. If the number of cross-sites is odd, a different cross-point is always assumed at the beginning of the string. The information (genes) between alternate pairs is exchanged as shown in Figure 3.4.

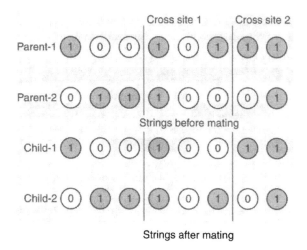

FIGURE 3.2 Two-point crossover.

3.1.2 FUNDAMENTAL DIFFERENCE

As seen from the above description of GAs working principles, GAs are very different from most traditional optimization methods. GAs work with a coding of variables instead of the variables themselves. The advantage of working with a coding of variable space is that the coding discretizes the search space even though the function may be continuous. Because function values at various discrete points are required, a discrete or discontinuous function may be tackled using GAs. This characteristic allows GAs to be applied to a wide variety of problem domains. Another advantage is that GA operators can exploit the similarities in string structures to make an effective search. More about this matter will be discussed a little later. One of the drawbacks of using a coding is that a proper coding of the problem must be used. The tentative guideline is that a coding that does not make the problem harder than the original problem must be used.

The more striking difference between GAs and most of the traditional optimization methods is that GAs work with a population of points instead of a single point.

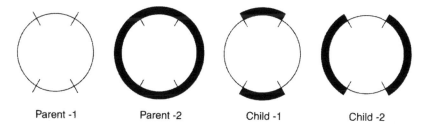

Parent -1   Parent -2   Child -1   Child -2

FIGURE 3.3 Multipoint crossover with odd number of cross sites.

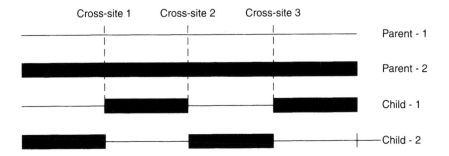

FIGURE 3.4 Multipoint crossover with even number of cross sites.

Because more than one string is processed simultaneously and used to update many strings in the population, the expected GA solution is very likely to be a global solution. Even though some traditional algorithms are population-based like Box's algorithms, those methods do not efficiently use the obtained information. Moreover, because a population is what is updated at every generation, a set of solutions (in the case of multimodal optimization, multi-objectives Pareto optimization, and others) can be obtained simultaneously.

In the prior discussion about GA operators or their working principles, nothing is mentioned about the gradient or any other auxiliary problem information. In fact, GAs do not require any auxiliary information except function values. The direct search methods used in traditional optimization also do not explicitly require gradient information but, in some of those methods, search directions are found using objective function values similar in concept to the gradient of the function. Moreover, some direct search methods work under the assumption that the function to be optimized is unimodal.

The other difference is that GAs use probabilistic rules to guide their search. This process may look *ad hoc* but careful consideration can provide some interesting properties of this type of search. The basic problem with most of the traditional methods is that fixed transition rules exist to move from one point to another point. This behavior is why those methods, in general, can only be applied to a special class of problems where any point in the search space leads to the desired optimum. Thus, these methods are not robust and simply cannot be applied to a wide variety of problems.

In trying to solve any other problem, if a mistake is made early, it becomes very hard to recover from that mistake because fixed rules are used. GAs, on the other hand, use probabilistic rules and an initial random population. Thus, the search can proceed in any direction early and no major decision is made in the beginning. Later, when the population has converged in some locations, the search direction narrows and a near-optimal solution is found. This nature of narrowing the search space as the generation progresses is adaptive and is a unique characteristic to genetic algorithms. This characteristic of GAs also permits them to be applied to a wide class of problems, giving them the robustness that is useful in the very sparse nature of engineering design problems.

Another difference with most of the traditional methods is that GAs can be easily and conveniently used in parallel machines. By using tournament selection — where two strings are picked at random and the better string is copied in the mating pool — instead of proportionate selection, only two processors are involved at a time. Because any crossover operator requires an interaction between only two strings and mutation requires alteration in only one string at a time, GAs are suitable for parallel machines. In real-world design optimization problems, most of the computational time is spent evaluating a solution; with multiple processors, all solutions in a population can be evaluated simultaneously. This advantage can reduce substantially the overall computational time.

Every good optimization method needs to balance the extent of exploration of the information obtained until the current time with the extent of exploitation of the search space required to obtain new and better points. If the solutions obtained are exploited too much, premature convergence is expected. If too much stress is given on searching, the information obtained thus far has not been used properly. Therefore, the solution time can be enormous and the search is similar to a random search method. Most traditional methods have fixed transition rules and thus have a fixed amount of exploration and exploitational consideration. For example, a pattern search algorithm has a local exploratory search (its extent is fixed beforehand) followed by a pattern search. The exploitation aspect comes only in the determination of search directions. Box's method has almost no exploration consideration and hence is not very effective. In contrast, the exploitation and exploration aspects of GAs can be controlled almost independently. This quality provides a lot of flexibility in designing a GA.

3.1.3 GA PARAMETERS

The building block hypothesis gives an intuitive and qualitative reasoning to what might cause GAs to work but it reveals nothing about for what values of various GA parameters it would work. In this subsection, some guidelines to determine values for GA parameters are presented.

The choice of the string length is the first decision to be made. The string length is usually chosen depending on the accuracy needed in the solution. For example, if binary strings of length 5 are used, the search space would contain 31 strings. The minimum accuracy in the solution that can be expected using GAs would be approximately 1/31 of the search space.

Apparently, the building block hypothesis suggests that problems that can be solved successfully using GAs must be linearly separable in terms of clusters of bits, but this is not the case. The above hypothesis can also be applied in problems having higher order nonlinearities with the one requirement: all competing building blocks are supplied in the initial population either by means of biased initial population or by making the initial random value large so that high-order scheme competitions can take place in the population. Messy GAs were developed to supply building blocks without stray bits and found to be successful in solving many difficult problems.

The other important issue is the balance between the exploitation and exploration aspects of GA operators. Reproduction is responsible for exploring the current population by making many duplicates of good strings, and crossover and mutation are responsible for exploring a set of good strings for better strings. GA success is, therefore, dependent on a nice balance between the two. If too many copies of the good strings are allocated in the mating pool, the diversity of the mating pool is reduced, which in turn reduces the extent of search that can be accomplished using crossover and mutation operators. Even though this aspect of GAs was discussed earlier to provide flexibility in their design, this quality could cause some potential problems if GA operators are not properly designed to have a proper balance between the two. Recently, a control map became available for values of the selection pressure, S (the number of copies allocated to the best string in the population), versus the crossover probability (the extent of search) for bit-wise linear problems using a computational model that equates the degree of characteristic time of convergence of the selection and the crossover operators alone.

3.1.4 SELECTION METHODS

After deciding on coding, the second decision to make in using a genetic algorithm is how to perform the selection — that is, how to chose the individuals in the population that will create offspring for the next generation and how many offspring each will create. The purpose of selection is, of course, to emphasize fitter individuals in the population, in hopes that their offspring will, in turn, have even higher fitness. Selection must be balanced with variation from crossover and mutation (the "exploitation/exploration balance"): too strong selection means that suboptimal, highly fit individuals will take over the population, reducing the diversity needed for further change and progress; too weak selection will result in too slow evolution. As was the case for encodings, numerous selection schemes have been proposed in the GA literature.

Some of the most common methods are described below. As was the case for encodings, these descriptions do not provide rigorous guidelines about which method to use for which problem; this is still an open question for GAs.

3.1.4.1 Fitness-Proportionate Selection with "Roulette Wheel" and "Stochastic Universal" Sampling

Holland's original GA used fitness-proportionate selection, in which the "expected value" of an individual (i.e., the expected number of times an individual will be selected to reproduce) is that individual's fitness divided by the average fitness of the population. The most common method for implementing this is roulette wheel sampling, described earlier: each individual is assigned a slice of a circular "roulette wheel," the size of the slice being proportional to the individual's fitness. The wheel is spun N times, where N is the number of individuals in the population. On each spin, the individual under the wheel's marker is selected to be in the pool of parents for the next generation. This method can be implemented as follows:

Add the total expected value of individuals in the population. Call this sum T.

Repeat N times.

Choose a random integer r between 0 and T.

Loop through the individuals in the population, summing the expected values, until the sum is greater than or equal to r. The individual whose expected value puts the sum over this limit is the one selected.

This stochastic method statistically results in the expected number of offspring for each individual. However, with the relatively small populations typically used in GAs, the actual number of offspring allocated to each individual is often far from its expected value (an extremely unlikely series of spins of the roulette wheel could even allocate all offspring to the worst individual in the population). James Baker (1987) proposed a different sampling method — stochastic universal sampling (SUS) — to minimize this "spreas" (the range of possible actual values, given an expected value). Rather than spin the roulette wheel N times to select N parents, SUS spins the wheel once — but with N equally spaced pointers that are used to select the N parents. Baker (1987) gives the following code fragment for SUS (in C):

```
ptr = Rand () ; //* Returns random number uniformly distributed in [0,1] *//
for (sum = i = 0 ; i < N; i++);
for (sum + = ExpVal (i,t); sum > ptr; ptr++);
select (i);
```

where i is an index over population members and where $ExpVal(i,t)$ gives the expected value of individual i at time t. Under this method, each individual i is guaranteed to reproduce at least $[ExpVal(i,t)]$ times but no more than $[ExpVal(i,t)]$ times.

SUS does not solve the major problems with fitness-proportionate selection. Typically, early in the search the fitness variance in the population is high and a small number of individuals are much fitter than the others. Under fitness-proportionate selection, they and their descendents will multiply quickly in the population, in effect preventing the GA from doing any further exploration. This is known as premature convergence. In other words, an early fitness-proportionate selection often puts too much emphasis on "exploitation" of highly-fit strings at the expense of explorating other regions of the search space. Later in the search, when all individuals in the population are very similar (the fitness variance is low), no real fitness differences exist for selection to exploit and evolution grinds to a near halt. Thus, the rate of evolution depends on the variance of fitnesses in the population.

3.1.4.2 Sigma Scaling

To address such problems, GA researchers have experimented with several "scaling" methods — methods for mapping "raw" fitness values to expected values so as to make the GA less susceptible to premature convergence. One example is sigma scaling, which keeps the selection pressure (i.e., the degree to which highly fit individuals are allowed many offspring) relatively constant over the course of the run rather than depending on the fitness variances in the population. Under sigma scaling, an individual's expected value is a function of its fitness, the population mean, and the population standard deviation. An example of sigma scaling would be

$$ExpVal(i,t) = \begin{cases} 1 + \dfrac{f(i) - \bar{f}(t)}{2\sigma(t)} & \text{if } \sigma(t) \neq 0 \\ \\ 1.0 & \text{if } \sigma(t) = 0 \end{cases}$$

where $ExpVal(i,t)$ is the expected value of individual i at time t, $f(i)$ is the fitness of i, $\bar{f}(t)$ is the mean fitness of the population at time t, and (t) is the standard deviation of the population fitnesses at time t. This function, used in the work of Tanese (1989), gives an individual with fitness one standard deviation above the mean 1.5 expected offspring. If $ExpVal(i,t)$ was less than 0, Tanese arbitrarily reset it to 0.1 so that individuals with very low fitness had some small chance of reproducing.

At the beginning of a run when the standard deviation of fitnesses is typically high, the fitter individuals will not be many standard deviations above the mean and so they will not be allocated the majority of offspring. Likewise, later in the run when the population is typically more converged and the standard deviation is lower, the fitter individuals will stand out more, allowing evolution to continue.

3.1.4.3 Elitism

Elitism, first introduced by Kenneth De Jong (1975), is an addition to many selection methods that forces the GA to retain some number of the best individuals at each generation. Such individuals can be lost if they are not selected to reproduce or if they are destroyed by crossover or mutation. Many researchers have found that elitism significantly improves the GA performance.

3.1.4.4 Boltzmann Selection

Sigma scaling keeps the selection pressure constant over a run. But often different amounts of selection pressure are needed at different times in a run — for example, early on it might be good to be liberal, allowing less fit individuals to reproduce at close to the rate of fitter individuals and having selection occur slowly while maintaining a lot of variation in the population. Having a stronger selection later on may be good in order to strongly emphasize highly fit individuals, assuming that the early diversity with slow selection has allowed the population to find the right part of the search space.

One approach to this procedure is Boltzmann selection (an approach similar to simulated annealing), in which a continuously varying "temperature" controls the rate of selection according to a preset schedule. The temperature starts out high, meaning that selection pressure is low (i.e., every individual has some reasonable probability of reproducing). The temperature is gradually lowered, gradually increasing the selection pressure and thereby allowing the GA to narrow in ever more closely to the best part of the search space, while maintaining the "appropriate" degree of diversity. A typical implementation is to assign to each individual i an expected value,

$$ExpVal(i,t) = \frac{e^{f(i)/T}}{\left\langle e^{f(i)/T} \right\rangle_t}$$

where T is temperature and the bracketed term denotes the average over the population at time t. Experimenting with this formula will show that as T decreases, the difference in $ExpVal(i,t)$ between high and low fitnesses increases. The desired condition is to have this happen gradually over the course of a search, so the temperature is gradually decreased according to a predefined schedule. It is found that (Michell, 1998) this method outperformed fitness-proportionate selection on a small set of test problems.

Fitness-proportionate selection is commonly used in GA mainly because it was part of Holland's original proposal and because it is used in the Schema theorem; but evidently, for many applications, simple fitness-proportionate selection requires several "fixes" to make it work well. In recent years, completely different approaches to selection (e.g., rank and tournament selection) have become increasingly common.

3.1.4.5 Rank Selection

Rank selection is an alternative method whose purpose is also to prevent too-quick convergence. In the version proposed by Baker (1985), the individuals in the population are ranked according to fitness and the expected value of each individual depends on its rank rather than on its absolute fitness. No need to scale fitnesses exists in this case because absolute differences in fitness are obscured. This discarding of absolute fitness information can have advantages (using absolute fitness can lead to convergence problems) and disadvantages (in some cases, knowing that one individual is far fitter than its nearest competitor might be important). Ranking avoids giving the far largest share of offspring to a small group of highly fit individuals and thus reduces the selection pressure when the fitness variance is high. It also keeps up selection pressure when the fitness variance is low; the ratio of expected values of individuals ranked i and $i + 1$ will be the same whether their absolute fitness differences are high or low.

The liner ranking method proposed by Baker is as follows: Each individual in the population is ranked in increasing order of fitness, from 1 to N. The user chooses the expected value of each individual with rank N and with $Max \geq 0$. The expected value of each individual i in the population at time t is given by

$$ExpVal\ (i,t) = Min + (Max - Min)\ \frac{rank(i,t) - 1}{N - 1}$$

where *Min* is the expected value of the individual with rank 1. Given the constraints *Max* ≥ 0 and $\sum_i$ *ExpVal* $(i,t) = N$ (since population size stays constant from generation to generation), it is required that $1 \le Max \le 2$ and $Min = 2 - Max$.

At each generation the individuals in the population are ranked and assigned expected values according to the above equation. Baker recommended *Max* = 1.1 and showed that this scheme compared favorably to fitness-proportionate selection on some selected test problems. Rank selection has a possible disadvantage: Slowing down selection pressure means that the GA will in some cases be slower in finding highly fit individuals. However, in many cases the increased preservation of diversity that results from ranking leads to more successful searches than the quick convergence that can result from fitness-proportionate selection. A variety of other ranking schemes (such as exponential rather than linear ranking) have also been tried. For any ranking method, once the expected values have been assigned, the SUS method can be used to sample the population (i.e., choose parents).

A variation of rank selection with elitism was used by some researchers. In those examples, the population was ranked by fitness and the top *E* strings were selected to be parents. The $N - E$ offspring were merged with the *E* parents to create the next population. As was mentioned above, this method is a form of the so-called $(\mu + \lambda)$ strategy used in the evolution strategies community. This method can be useful in cases where the fitness function is noisy (i.e., a random variable, possibly returning different values on different calls on the same individual); the best individuals are retained so they can be tested again and thus, over time, gain increasingly reliable fitness estimates.

3.1.4.6 Tournament Selection

The fitness-proportionate methods described above require two passes through the population at each generation: one pass to compute the mean fitness (and, for sigma scaling, the standard deviation) and one pass to compute the expected value of each individual. Rank scaling requires sorting the entire population by rank — a potentially time-consuming procedure. Tournament selection is similar to rank selection in terms of selection pressure, but it is computationally more efficient and more amenable to parallel implementation. Two individuals are chosen at random from the population. A random number *r* is then chosen between 0 and 1. If r < k (where k is a parameter, for example 0.75), the fitter of the two individuals is selected to be a parent; otherwise the less fit individual is selected. The two are then returned to the original population and can be selected again. An analysis of this method was presented by Goldberg and Deb (1991).

3.1.4.7 Steady-State Selection

Most GAs described in the literature have been "generational" — at each generation, the new population consists entirely of offspring formed by parents in the previous generation (though some of these offspring may be identical to their parents). In some schemes, such as the elitist schemes described above, successive generations overlap to some degree — some portion of the previous generation is retained in the new population. The fraction of new individuals at each generation has been called the "generation gap" (De Jong, 1975). In steady-state selection, only a few individuals are replaced in each generation: usually a small number of the least fit individuals are replaced by offspring resulting from crossover and mutation of the fittest individuals. Steady-state GAs are often used in evolving rule-based systems (e.g., classifier systems), in which incremental learning (and remembering what has already been learned) is important and in which members of the population collectively (rather than individually) solve the problem at hand.

3.1.5 INHERITANCE OPERATORS

Some of the inheritance operators (low level operators) used in genetic algorithms:

Inversion
Dominance
Deletion
Intrachromosomal duplication
Translocation
Segregation
Specification
Migration
Sharing
Mating

3.1.6 MATRIX CROSSOVER (TWO-DIMENSIONAL CROSSOVER)

Normally, the strings are represented as a single dimensional array as shown in Figure 3.5. In the above case, two strings of length 4 are concatenated to form an individual. The cross-site selected for this case is obviously single-dimensional; whereas in the case of two-dimensional crossover, each individual is represented as a two-dimensional array of vectors to facilitate the process. The process of two-dimensional crossover is depicted in Figure 3.6.

Two random sites along each row and column are chosen, then the string is divided into nonoverlapping rectangular regions. Two cross-sites, both row- and column-wise, will divide each individual into nine overlapping rectangular regions, three layers horizontally and vertically. Select any region in each layer,

String -1	1 0 1 1			1 0 0 1	
String -2	0 1 0 1			1 1 1 0	
	Substring -1			Substring -2	

FIGURE 3.5 Single dimensional strings.

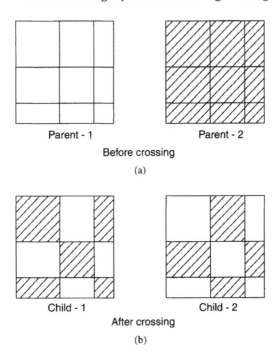

FIGURE 3.6 (a) Matrix crossover — before crossing. (b) Matrix crossover — after crossing.

either vertically or horizontally, and then exchange the information in that region between the matted populations. The selection of crossover operator is made so that the search in genetic space is proper. In the case of the single point crossover operator, the search is not extensive, but maximum information is preserved between parents and children.

3.1.7 Inversion and Deletion

3.1.7.1 Inversion

A string from the population is selected and the bits between two random sites are inverted, as shown in Figure 3.7.

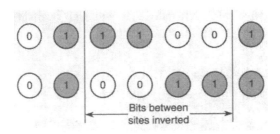

FIGURE 3.7 Inversion.

3.1.7.2 Linear + End-Inversion

Linear + end-inversion performs linear inversion with a specified probability of 0.75. If linear inversion was not performed, the end inversion would be performed with equal probability of 0.125 at either the left or right end of the string. Under end inversion, the left or right end of the string was picked as one inversion-point and a second inversion-point was picked uniformly at random from the point no farther away than one half of the string length. Linear + end-inversion minimizes the tendency of linear inversion to disrupt bits located near the center of the string disproportionately to those bits located near the ends.

3.1.7.3 Continuous Inversion

In continuous inversion, inversion is applied with specified inversion probability P_r to each new individual when it is created.

3.1.7.4 Mass Inversion

No inversion takes place until a new population is created and, thereafter, one half of the population undergoes identical inversion (using the same two inverting points).

3.1.7.5 Deletion and Duplication

Any two or three bits in random order are selected and the previous bits are duplicated, as shown in Figure 3.8.

3.1.8 Crossover and Inversion

The crossover and inversion operator is the combination of both crossover and inversion operators. In this method, two random sites are chosen, the contents bracketed by these sites are exchanged between two mated parents, and the end points of these exchanged contents switch place. For example, if the cross-sites in parents shown in Figure 3.9 are 2 and 7, the crossover and inversion operation is performed in the way shown.

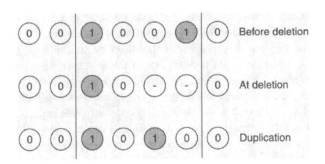

FIGURE 3.8 Deletion and duplication.

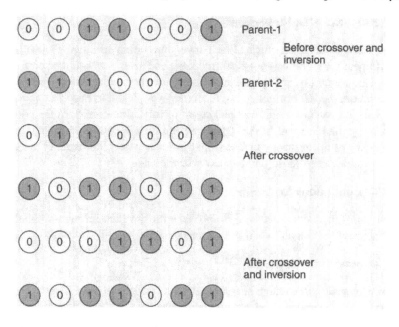

FIGURE 3.9 Crossover and inversion.

3.2 SIMULATED ANNEALING (SA)

The simulated annealing method resembles the cooling process of molten metals through annealing. At high temperatures, the atoms in the molten metal can move freely with respect to each other; but as the temperature is reduced, the movement of the atoms becomes restricted. The atoms start to order themselves and finally form crystals having a minimum possible energy. However, the formation of the crystal mostly depends on the cooling rate. If the temperature is reduced at a very fast rate, the crystalline state might not be achieved at all; instead, the system can end up in a polycrystalline state. Therefore, to achieve the absolute minimum energy state, the temperature must be reduced at a slow rate. The process of slow cooling is known as annealing in metallurgy.

The simulated annealing procedure simulates this process of annealing to achieve the minimum function value in a minimization problem. The slow cooling phenomenon of the annealing process is simulated by controlling a temperature-like parameter introduced with the concept of the Boltzmann probability distribution. According to the Boltzmann probability distribution, a system in thermal equilibrium at a temperature T has its energy distributed probabilistically according to $P(E) = e^{E/kT}$, where k is the Boltzmann constant. This expression suggests that a system at a high temperature has an almost uniform probability of being at any energy state; but at a low temperature, it has a small probability of being at a high energy state. Therefore, by controlling the temperature T and assuming that the search process follows the Boltzmann probability distribution, the convergence of an algorithm can be controlled.

Metropolis et al. (1953) suggested one way to implement the Boltzmann probability distribution in simulated thermodynamic systems. The same can be found in the function minimization context. At any instant, the current point is x_t and the function value at that point is $E(t) = f(x_t)$. Using Metropolis' algorithm, the probability of the next point being at $x_t + 1$ depends on the difference in the function values at these two points or on $\Delta E = E(t + 1) - E(t)$ and is calculated using the Boltzmann probability distribution:

$$P(E(t + 1)) = min\ [1,\ e^{-\Delta E/kT}].$$

If $\Delta E \geq 0$, this probability is 1 and the point $x_t + 1$ is always accepted. In the function minimization context, this result makes sense because if the function value at $x_t + 1$ is better than at x_t, the point $x_t + 1$ must be accepted. The interesting situation happens when $\Delta E > 0$, which implies the function value at $x_t + 1$ is worse than at x_t. According to Metropolis' algorithm, some finite probability of selecting the point $x_t + 1$ exists even though it is worse than the point x_t. However, this probability is not the same in all situations. This probability depends on the relative magnitude of the ΔE and T values. If the parameter T is large, this probability is greater for points with largely disparate function values. Thus, any point is almost acceptable for a larger value of T. If the parameter T is small, the probability of accepting an arbitrary point is small. Thus, for small values of T, the points with only small deviation in function value are accepted.

The above procedure can be used in the function minimization of certain cost functions. The algorithm begins with an initial point x_1 and a high temperature T. A second point x_2 is created at random in the vicinity of the initial point and the difference in the function values (ΔE) at these two points is calculated. If the second point has a smaller function value, the point is accepted; otherwise, the point is accepted with a probability $e^{-\Delta E/T}$. This completes one iteration of the simulated annealing procedure. In the next generation, another point is created at random in the neighborhood of the current point and the Metropolis algorithm is used to accept or reject the point. To simulate the thermal equilibrium at every temperature, a number of points is usually tested at a particular temperature before reducing the temperature. The algorithm is terminated when a sufficiently small temperature is obtained or a small enough change in function values is found.

3.2.1 Optimization Procedure Using SA

Step 1: Choose an initial point, x_1, and a termination criteria. Set T to a sufficiently high value, the number of iterations to be performed at a particular temperature n, and set $t = 0$.

Step 2: Calculate a neighboring point x_2. Usually, a random point in the neighborhood is created.

Step 3: If $\Delta E = E(x_2) - E(x_1) < 0$, set $t = t + 1$; else create a random number r in the range $(0,1)$. If $r \leq e^{-\Delta E/KT}$, set $t = t + 1$; else go to Step 2.

Step 4: If $x_2 - x_1 < T$ and T is small, terminate; else if $(t \bmod n) = 0$, lower T according to a cooling schedule. Go to Step 2.

3.3　ANT COLONY OPTIMIZATION (ACO)

Ant colony optimization is a metaheuristic approach to tackling a hard CO problem that was first proposed in the early 1990s by Dorigo, Maniezzo and Colorni. Fascinated by the ability of the almost blind ants to establish the shortest route from their nests to the food source and back, researchers found that these ants secrete a substance called pheromones and use its trails as a medium for communicating information among themselves. Also, they are capable of adapting to changes in the environment, such as finding a new shortest path when the old one is no longer available due to a new obstacle. In Figure 3.10, ants are moving on a straight line that connects a food source to their nest. The primary, well-known means for ants to form and maintain the line is a pheromone trail. Ants deposit a certain amount of pheromone while walking and each ant probabilistically prefers to follow a direction rich in pheromone. This elementary behavior of real ants can explain how they can find the shortest path that reconnects a broken line after the sudden appearance of an unexpected obstacle in the initial path (Figure 3.11). Once the obstacle has appeared, ants right in front of the obstacle cannot continue to follow the pheromone trail in the straight line. In this situation, some ants choose to turn right and others choose to turn left (Figure 3.12).

The ants that choose, by chance, the shorter path around the obstacle will more rapidly reconstitute the interrupted pheromone trail compared to those that choose the longer path. Thus, the shorter path will receive a greater amount of pheromone per time unit and, in turn, a larger number of ants will choose the shorter path. Due to this positive feedback (autocatalytic) process, all the ants will rapidly choose the shorter path (Figure 3.13). The most interesting aspect of this autocatalytic process is that finding the shortest path around the obstacle seems to be an emergent property of the interaction between the obstacle shape and ants' distributed behavior.

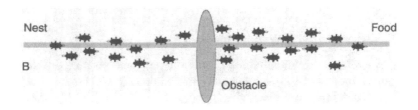

FIGURE 3.10 Real ants follow a path between nest and food source.

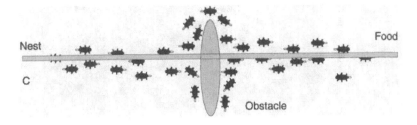

FIGURE 3.11 An obstacle appears on the path: Ants choose whether to turn left or right.

Although all ants move at approximately the same speed and deposit a pheromone trail at approximately the same rate, the fact that it takes longer to contour obstacles on their longer side than on their shorter side makes the pheromone trail accumulate more quickly on the shorter side. The ants's preference for higher pheromone trail levels makes this accumulation even quicker on the shorter path. A similar process used in a simulated world inhabited by artificial ants can solve a hard CO problem.

The artificial ants used to mimic the behavior of real ants in ACO differ in a few respects:

These ants are not completely blind.
They ants have some memory.
They live in an environment where time is discrete.

3.3.1 STATE TRANSITION RULE

Starting in an initial node, every ant chooses the next node in its path according to the state transition rule,

$$P_{ij}^{k} = \frac{(\tau_{ij})^{\alpha}(\eta)^{\beta}}{\sum_{k=1}^{n}(\tau_{ij})^{\alpha}(\eta)^{\beta}}$$

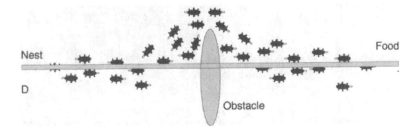

FIGURE 3.12 Pheromone is deposited more quickly on the shorter path.

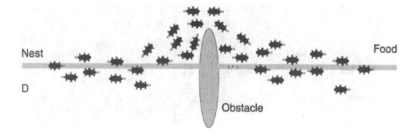

FIGURE 3.13 All ants have chosen the shorter path.

where

P_{ij}^k = probability that ant k will choose the next node j
n = number of jobs
$_{ij}$ = pheromone trial
= heuristic distance (processing time of the job)

After choosing the node, the pheromone value is updated using the pheromone updating rule.

3.3.2 PHEROMONE UPDATING RULE

Pheromone updating follows three different rules. They are:

Ant cycle
Ant quantity
Ant density

Experiments conducted in ant colony optimization by Marco Dorigo et al. show that the ant cycle pheromone update rule is best compared to others. The ant cycle pheromone update rule is used thus:

$$\tau(t + 2) = \rho.\tau + \sum_{k=1}^{n} \Delta\tau_{ij}(t)$$

$$\Delta\tau_{ij} = 1/c_0^k$$

where

$\Delta\tau_{ij}$ = increment in the pheromone level
c_0^k = combined objective function value for ant k
ρ = evaporation coefficient

This rule consists of two actions. First, a fraction of pheromone on all edges is evaporated. Second, an increment of pheromone is given to those edges that are scheduled within the solution of the ant that so far has the best solution to the problem.

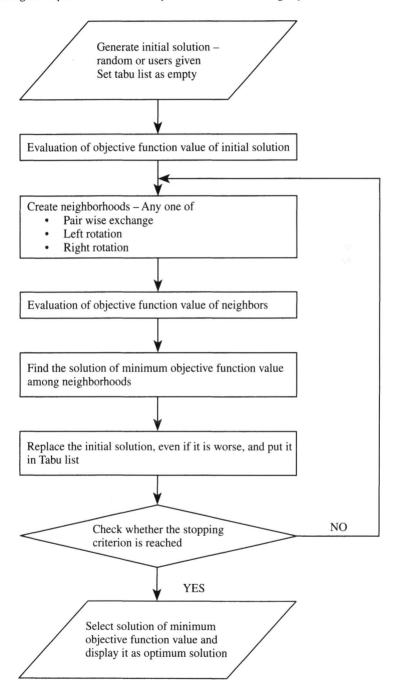

FIGURE 3.14 Tabu search algorithm.

In the next cycle, those edges belonging to the best solution will have a higher probability, thus exploring this information performs a certain kind of reinforcement learning. On the other hand, evaporation prevents searching in the neighborhood of a local minimum. The evaluation of the algorithm is meant to cyclically minimize the evaluation function.

3.3.3 STEPS IN ANT COLONY ALGORITHM

Step 1: Intialization
   Set $t = 0$; $NC = 0$; [t is the time counter, NC is the number of algorithm cycles]
   For each edge (i, j), set an initial value $\tau_{1j} = c$, $\Delta x_{1j} = 0$.
   [$\tau_{1j}(t)$ is the intensity of trial on the edge (i,j) at time t]
   [Δx_{1j} is the quantity of trial laid on edge (i,j) by the kth ant]

Step 2: Starting node
   For each ant k, place ant k on the randomly chosen node and store this information in Tabu$_k$.

Step 3: Build a tour for each ant
   For each ant i and for each ant k, choose the node j to move to with the probability $P_{ij}^k(t)$ given by the formula.
   Store the information in the Tabu$_k$.

Step 4: Update the intensity of trial
   For each ant k, compute the quantities Δx_{ij} laid on each edge (i,j) according to Q/L_k.
   For each edge (Δx_{ij}), compute the intensity of trial according to the equation
   $\tau_{1j}(t + 1) = p.\Delta x_{1j}(t) + \xi \, \Delta x_{1j}^k$.
   $T = t + 1$
   $NC = NC + 1$
   $\Delta x_{1j}^k = 0$.

Step 5: Termination condition
   Memorize the shortest tour found to this point.
   If $NC < NC_{MAX}$ and is not stagnation behavior, empty all Tabu lists and go to Step 2; else stop.

3.4 PARTICLE SWARM OPTIMIZATION (PSO)

Particle swarm optimization is a population-based stochastic optimization technique developed by Eberhart and Kennedy in 1995, inspired by the social behavior of bird flocking or fish schooling.

3.4.1 BACKGROUND OF ARTIFICIAL LIFE

The term "artificial life" (Alife) is used to describe research into human-made systems that possess some of the essential properties of life. Alife includes the two-fold research topic:

> Alife studies how computational techniques can help when studying biological phenomena.
> Alife studies how biological techniques can help with computational problems.

The focus of particle swarm optimization is on the second Alife topic. Actually, many computational techniques have already been inspired by biological systems. For example, an artificial neural network is a simplified model of the human brain and the genetic algorithm is inspired by natural selection.

Here we discuss another type of biological system — a social system, more specifically, the collective behaviors of simple individuals interacting with their environment and each other, sometimes called "swarm intelligence." Two popular swarm-inspired methods exist in computational intelligence areas: ant colony optimization (ACO) and particle swarm optimization (PSO). ACO was inspired by the behaviors of ants and has many successful applications in discrete optimization problems.

3.4.2 PARTICLE SWARM OPTIMIZATION TECHNIQUE

The particle swarm concept originated as a simulation of a simplified social system. The original intent was to graphically simulate the choreography of a bird flock or fish school. However, the particle swarm model can be used as an optimizer.

As stated previously, PSO simulates the behaviors of bird flocking. Suppose a group of birds are randomly searching for food in an area. Only one piece of food exists in the area being searched. All the birds do not know where the food is, but they know how far the food is in each iteration. So what is the best strategy to find the food? The most effective one is to follow the bird nearest to the food.

PSO learns from this scenario and uses it to solve optimization problems. In PSO, each single solution is a "bird" in the search space (we call it a "particle"). All particles have fitness values that are evaluated by the fitness function to be optimized and have velocities that direct the "flying" of the particles. The particles fly through the problem space by following the current optimum particles.

PSO is initialized with a group of random particles (solutions) and then searches for optima by updating generations. In every iteration, each particle is updated by following two "best" values. The first one is the best solution (fitness) it has achieved so far. The fitness value is also stored; this value is called *pbest*. Another "best" value tracked by the particle swarm optimizer is the best value obtained so far by any particle in the population. This best value is a global best and is called *gbest*. When a particle takes part of the population as its topological neighbors, the best value is a local best and is called *lbest*.

After finding the two best values, the particle updates its velocity and positions. Eberhart and Shi (2000) introduced an inertia weight factor that dynamically adjusted the velocity over time, gradually focusing the PSO into a local search. The particle updates its velocity and positions with the following: Equation (3.1) and Equation (3.2).

$$v[\ ] = \omega\, v[\ ] + c_1\, rand(\ ) \times (pbest[\ ] - present[\ ]) + c_2\, rand(\ ) \times (gbest[\ ] - present[\ ])$$

$$(3.1)$$

$$present[\ ] = present[\ ] + v[\ ] \qquad (3.2)$$

where

$v[\ ]$ = particle velocity
$present[\ ]$ = current particle (solution)
$pbest[\ ]$ = best solution among each particle
$gbest[\ ]$ = best among defined as stated before
$rand(\ )$ = random numbers between (0,1)
ω = inertia weights, usually 0.8 or 0.9
c_1, c_2 are learning factors. Usually, $c_1 = c_2 = 2$.

3.4.3 ALGORITHM OF PARTICLE SWARM OPTIMIZATION

Most evolutionary techniques have the following procedure:

Random generation of an initial population.
Reckoning of a fitness value for each subject, depending directly on the distance to the optimum.
Reproduction of the population based on fitness values.
If requirements are met, then stop. Otherwise, go back to Step 2.

From the procedure, observe that PSO shares many common points with GA. Both algorithms start with a group of a randomly generated population and both have fitness values to evaluate the population. Both update the population and search for the optimum with random techniques. Both systems do not guarantee success. However, PSO does not have genetic operators like crossover and mutation. Particles update themselves with the internal velocity. They also have memory, which is important to the algorithm.

3.4.4 PSO PARAMETERS CONTROL

As the previous case shows, two key steps exist when applying PSO to optimization problems: the representation of the solution and the fitness function. One of the advantages of PSO is that it takes real numbers as particles, unlike GA that must change to binary encoding unless special genetic operators are used. Then the standard procedure can be used to find the optimum. The searching is a repeat process and the stop criteria are that the maximum iteration number is reached or the minimum error condition is satisfied.

Not many parameters need to be tuned in PSO. A list of the parameters and their typical values is presented below.

Number of particles: The typical range is 20 to 40. Actually, for most of the problems, 10 particles is large enough to get good results. For some difficult or special problems, one can try 100 or 200 particles as well.

Dimension of particles: This quantity is determined by the problem to be optimized.

Range of particles: This value is also determined by the problem to be optimized. Different ranges for different dimensions of particles can be specified.

V_{max}: This value determines the maximum change one particle can take during one iteration. The range of the particle is usually set as the V_{max}.

Learning factors: c_1 and c_2 are usually equal to 2. However, other settings are also used in different papers but generally $c_1 = c_2$ and with a range [0, 4].

Stop condition: This value is the maximum number of iterations the PSO executes and the minimum error requirement. The maximum number of iterations is set to 2000. This stop condition depends on the problem to be optimized particularly with respect to the search space.

Global version versus local version: Two versions of PSO were introduced, a global and a local version. The global version is faster but might converge to a local optimum for some problems. The local version is a little bit slower but is not easily trapped into local optima. Use the global version to get quick results and the local version to refine the search.

3.4.5 COMPARISONS BETWEEN GENETIC ALGORITHM AND PSO

Compared with genetic algorithms, the information sharing mechanism in PSO is significantly different. In GA, since chromosomes share information with each other, the whole population moves like a single group toward an optimal area. In PSO only *gbest* (or *lbest*) gives information to others in a one-way information sharing mechanism. Evolution only looks for the best solution. Compared with GA, all the particles tend to quickly converge to the best solution, even in the local version in most cases.

PSO shares many similarities with evolutionary computation techniques such as GA. The system is initialized with a population of random solutions and searches for optima by updating generations. However, unlike GA, PSO has no evolution operators, such as crossover and mutation. In PSO, the potential solutions fly through the problem space by following the current optimum particles.

Compared to GA, the advantages of PSO are that it is easy to implement and has few parameters to adjust. PSO has been successfully applied in many areas: function optimization, artificial neural network training, fuzzy system control, and other areas where GA can be applied.

3.5 TABU SEARCH (TS)

The Tabu search is an iterative procedure designed for the solution of optimization problems. TS was invented by Glover and has been used to solve a wide range of hard optimization problems, such as job shop scheduling, graph coloring, the traveling salesman problem (TSP), and the capacitated arc routing problem.

The basic concept of a Tabu search as described by Glover (1986) is "a meta-heuristic superimposed on another heuristic. The overall approach is to avoid entrainment in cycles by forbidding or penalizing moves that take the solution, in the next iteration, to points in the solution space previously visited (hence 'tabu')." The Tabu search is fairly new; Glover attributes its origin to about 1977. The method is still actively researched and continues to evolve and improve. The Tabu method was partly motivated by the observation that human behavior appears to operate with a random element that leads to inconsistent behavior, given similar circumstances.

As Glover points out, the resulting tendency to deviate from a charted course might be regretted as a source of error, but can also prove to be an advantage. The Tabu method operates in this way with the exception that new courses are not chosen randomly. Instead, the Tabu search proceeds according to the supposition that no advantage exists in accepting a new (poor) solution unless it is to avoid a path already investigated. This ensures that new regions of a problem's solution space will be investigated with the goal of avoiding local minima and ultimately finding the desired solution.

The Tabu search begins by marching to local minima. To avoid retracing the steps used, the method records recent moves in one or more Tabu lists. The original intent of the list was not to prevent a previous move from being repeated, but rather to ensure that it was not reversed. The Tabu lists are historical in nature and form the Tabu search memory. The role of the memory can change as the algorithm proceeds. At initialization, the goal is make a coarse examination of the solution space, known as "diversification;" but as candidate locations are identified the search is increasingly focused to produce local optimal solutions in a process of "intensification." In many cases, the differences between the various implementations of the Tabu method have to do with the size, variability, and adaptability of the Tabu memory to a particular problem domain.

3.5.1 TABU SEARCH ALGORITHM

TS is a local search procedure. Local search procedures can be compared on the following four design criteria:

The schedule representation needed for the procedure
The neighborhood design
The search process within the neighborhood
The acceptance-rejection criterion

3.5.2 GENERAL STRUCTURE OF TABU SEARCH

3.5.2.1 Efficient Use of Memory

The use of memory is an essential feature of TS; the Tabu conditions can usually be considered as a short-term memory that prevents cycling to some extent. Some efficient policies for the management of the Tabu lists are described below.

The use of memory may help intensify the search in "good" regions or diversify the search to unexplored regions.

3.5.3 Variable Tabu List Size

The basic role of the Tabu list is to prevent cycling. If the length of the list is too small, this role might not be achieved; conversely, a size that is too long creates too many restrictions and the mean value of the visited solutions grows with the increase of the Tabu list size. Usually, an order of magnitude of this size can be easily determined. However, given an optimization problem, finding a value that prevents cycling and that does not excessively restrict the search for all instances of the problem of a given size is often difficult or even impossible.

An effective way of circumventing this difficulty is to use a Tabu list with a variable size. Each element of the list belongs to it for a number of iterations that is bounded by given maximal and minimal values.

3.5.4 Intensification of Search

To intensify the search in promising regions, the search first must come back to one of the best solutions found so far. The size of the Tabu list then can be simply decreased for a "small" number of iterations. In some cases, more elaborate techniques can be used. Some optimization problems can be partitioned into subproblems. Solving these subproblems optimally and combining the partial solutions leads to an optimal solution. The difficulty with such a strategy obviously consists in finding a good partition.

3.5.5 Diversification

To avoid the circumstance of a large region of the state space graph remaining completely unexplored, diversifying the search is important. The simplest way to diversify is to perform several random restarts. A different way that guarantees the exploration of unvisited regions is to penalize frequently performed moves or solutions often visited. This penalty is set large enough to ensure the escape from the current region. The modified objective function is used for a given number of iterations. Using a penalty on frequently performed moves also is possible during the whole search procedure.

3.5.6 Stopping Criterion

If the maximum iteration number is reached or no improvements in the solution value are found for a fixed number of iterations, further iteration is stopped to help intensify the search in "good" regions or to diversify the search toward unexplored regions.

All the above intelligent techniques have been implemented for solving different manufacturing optimization problems and are described with numeric examples in the remaining chapters.

REFERENCES

Baker, J.E, Adaptive selection methods for genetic algorithms: Proceedings of the first international conference on genetic algorithms and their applications, Erlbaum, 1985.

Baker, J.E, Reducing bias and inefficiency in the selection algorithm, Genetic algorithms and their applications: Proceedings of the second international conference on genetic algorithm, Erlbaum, 1987.

Deb, K., *Optimization for Engineering Design — Algorithms and Examples*, Prentice Hall of India (Pvt.) Ltd., New Delhi, 1998.

De Jong, K., An analysis of the behaviour of a class of genetic adaptive systems, Ph.D. Thesis, University of Michigan, Ann Arbor, 1975.

Dorigo et al., The ant system: optimization by a colony of cooperating agents, *IEEE Transactions on Systems Man and Cybernetics,* B 26(1), 1–13, 1996.

Eberhart, R. and Kennedy, J., A new optimizer using particle swarm theory, *Proceedings of the Sixth International Symposium on Micro Machine and Human Science* (Nagoya, Japan), IEEE Service Center, Piscataway, NJ, 39–43, 1995.

Eberhart, R.C. and Shi, Y., Comparing inertia weights and constriction factors in particle swarm optimization, *Congress on Evolutionary Computing,* 1, 84–88, 2000.

Glover, F., Future paths for integer programming and links to artificial intelligence, Computers and Operations Research, 19, pp. 533–549, 1986.

Glover, F., Tabu search, part II, *ORSA Journal of Computers,* 2, 4–32, 1990.

Goldberg, D.E., *Genetic Algorithm in Search, Optimization and Machine Learning,* Addision Wesley, Singapore, 2000.

Goldberg, D.E., and Deb, K., A comparative analysis of selection schemes used in genetic algorithm, *Proceedings of Foundations of Gas and Classifier Systems,* Indiana University, Bloomington, IN, 69–74, 1990.

Goldberg, D.E., and Deb, K., A comparative analysis of selection schemes used in genetic algorithm, *Foundations of genetic algorithms, Morgan Kaufmann,* 1991.

Holland, John, Genetic algorithms and the optimal allocation of trials, *SIAM Journal of Computers,* 2, 88–105, 1974.

Jayaraman, V.K., et al., Ant colony framework for optimal design and scheduling of batch plants, *International Journal of Computers and Chemical Engineering,* 24, 1901–1912, 2000.

Kennedy, J. et al., Particle swarm optimization, *Proceedings of the IEEE International Conference on Neural Networks,* IV, IEEE Service Center, Piscatway, NJ, 1942–1948, 1995.

Maniezzo, V. et al., The ant system applied to the quadratic assignment problem, *IEEE Transactions on Knowledge and Data Engineering,* 11(5), 769–778, 1999.

Metropolis, N., Rosenbluth, A., Rosenbluth, M., Teller, A., and Teller, E., Equation of state calculations by fast computing machines, *Journal of Chemical Physics*, 21, pp. 1087–1092, 1953.

Mitchell, M., *An Introduction to Genetic Algorithms,* Prentice Hall of India (Pvt.) Ltd., New Delhi, 1998.

Proceedings of two-week short term training program, "Modeling and optimization of manufacturing systems using conventional and non-conventional techniques," J.J. College of Engineering and Technology, Trichirapalli, India, 2003.

Rajasekaran, S. and Vijayalakshmi Pai, G.A., *Neural Networks, Fuzzy Logic and Genetic Algorithms Synthesis and Applications,* Prentice Hall of India (Pvt.) Ltd., New Delhi, 2004.

Tanese, R., Distributed genetic algorithms for function optimization, Ph.D Thesis, Electrical Engineering and Computer Science Department, University of Michigan, 1989.

4 Optimal Design of Mechanical Elements

4.1 INTRODUCTION

A machine is a special type of mechanical structure characterized by mechanical elements constructed for the purpose of transmitting force or performing some physical function. Therefore, a machine design involves careful analysis and design of each and every mechanical element.

The design of mechanical elements is considerably more difficult than the problem of mechanical analysis because generally, for the latter, both geometry and materials are assumed to be constants and limitations such as space restrictions are not significant. The design of mechanical elements is always based on the satisfaction of mathematical equations, graphical information and tabulated data. The graphical information and tabulated data can be conveniently expressed as design equations.

Thus, design equations generally express functional requirements in terms of parameters that can be classified according to three basic groups:

Functional requirements parameters
Material parameters
Geometrical parameters

Functional requirements are the conditions that must be satisfied by the mechanical element for the structure or machine to work or function properly. Functional requirements can be positive and generally specified or generally implied, i.e., negative. Negative functional requirements are undesired effects whereas positive functional requirements are really desired effects.

The functional requirement parameters in a design equation are generally specified values from an analysis of the entire mechanical structure or machine before the actual design of the mechanical elements. Functional requirement parameters are primarily influenced by the factors external to the element and, depending on the particular machine structure, these parameters can possibly be dependent upon each other. So the functional requirement parameter group in a design equation is independent of the element being designed.

The material parameters in a design equation are generally not independent of each other. For most cases, material parameters cannot be changed individually in any arbitrary manner. Changing the materials can alter the value for the material parameter group. The material parameter group is independent of functional and geometrical parameters in a typical mechanical element design equation.

For the conduct of design studies, selecting independent geometrical parameters that define the geometry of the mechanical element uniquely is always possible and desirable. Redundancy in defining the geometry of the element should be avoided in the design procedure. In selecting the particular independent geometrical parameter, choosing parameters that are restricted to standard sizes or whose limits are either known or specified by the functional requirements of the entire mechanical structure or machine is wise.

4.1.1 ADEQUATE DESIGN

Adequate design can be defined as the selection of the material and the values for the independent geometrical parameters or the mechanical element so that the element satisfies its functional requirements and undesirable effects are kept to tolerable magnitudes.

Adequate design is often characterized by a "cut-and-try" method because of the existence of a rather loosely defined overall objective that results in an infinite number of possible design solutions. For many mechanical elements, adequate design is really the optimal design. This condition is true for cases where the undesirable effects are close to their tolerable limits by the application of cut-and-try.

In many cases, an optimal design study of the mechanical element will result in invaluable savings or in an appreciable improvement in product performance or quality. Also, many situations exist where adequate design would not readily reveal a satisfactory solution to a design problem because of certain practical limits and the erroneous conclusion of impossibility might be drawn. For such cases, the method of optimal design might more readily provide the possible solutions to the problem.

4.1.2 OPTIMAL DESIGN

Optimal design of a mechanical element is the selection of the material and the values for the independent geometrical parameters with the objective of either minimizing an undesirable effect or maximizing the functional requirement.

4.1.3 PRIMARY DESIGN EQUATION

In the optimal design of a mechanical element, the most important design equation is the one that expresses the quantity upon which the particular optimal design is based because it controls the procedure of design. For any particular mechanical element, the particular primary design equation will be determined by the most significant functional requirement or by the most significant undesirable effect.

4.1.4 SUBSIDIARY DESIGN EQUATIONS

In the optimal design of mechanical elements, the design equations other than the primary design equations are called the subsidiary design equations and generally express either functional requirements or significant undesirable effects whether they are directly specified or indirectly implied.

4.1.5 Limit Equations

Limitations on geometry are imposed by certain functional requirements of the mechanical structure or machine, such as space restrictions, practical manufacturing limitations in conjunction with material characteristics, and availability of standard sizes.

4.1.6 Optimal Design Procedure

For normal specifications, it is possible to develop the primary design equation to include the effects of all subsidiary equations and to extract the unique optimal design, while taking into account all significant specifications and limits.

The suggested optimal design procedure is outlined in the following steps:

Draw a freehand sketch of the mechanical element showing the significant basic geometry. Select the independent geometry parameters that will be used for uniquely defining the geometry of the element. If the choice exists, select geometrical parameters whose values are either specified as functional requirements or limited to permissible ranges.

Write the primary design equation that expresses the optimal design quantity. If possible, write this equation in terms of functional requirements, material parameters, and geometrical parameters.

Write all subsidiary design equations that express functional requirements and significant desirable effects.

Write all limit equations for functional requirements, significant undesirable effect parameters, and material and geometrical parameters.

Combine all subsidiary design equations with the primary design equation by eliminating an unlimited and unspecified common parameter from the primary equation for each subsidiary design equation. Doing this will so develop the primary design equation so that it consists only of specified values, independent parameters, and independent parameter groups.

Using the developed primary design equation from Step 5, roughly determine the variation of the optimal design quantity with respect to each independent parameter or independent parameter group in the primary design equation.

Apply the optimization algorithm for the selected independent parameters (mostly geometrical parameters) to get different values within the tolerable limits and determine the optimal values for the different parameters.

There is a lot of scope for formulating the conventional design procedures as optimization problems in order to improve the quality and other performance of the engineering products. Several non-conventional algorithms are available for solving different design optimization problems. In order to demonstrate the above, in this section genetic algorithm and simulated annealing techniques are described for solving the following design optimization problems:

Gear design
Three-bar truss design
Spring design
Single-point cutting tool design

4.2 GEAR DESIGN OPTIMIZATION

4.2.1 MATHEMATICAL MODEL OF GEAR DESIGN

4.2.1.1 Preliminary Gear Considerations

The first problem of the gear design is to find a design that can carry the power required. The gears must be big enough, hard enough, and accurate enough to do the job required. Finding out how much power a gear set must carry is frequently difficult. In some cases, loading will be continuous and in some cases, it will be an intermittently applied load, such as in the case of an internal combustion (IC) engine.

The following are the input parameters required: power to be transmitted (P) in kW; speed of the pinion (N) in rpm; and gear ratio (i). All three parameters are highly sensitive. Both driven and driver must be checked before these parameters are assigned a value. The quality of the gear and its maximum pitch line velocity, V_{max}, are selected from a database depending on the application. Suitable material for the gear and its limiting stress values are selected depending on the input parameters.

Objective function = [(Minimize error, weight), (Maximize efficiency)]

Since variation of error and weight is higher compared to efficiency, the importance of the objective is given accordingly (i.e., 40% to error, 40% to weight, and 20% to efficiency). Therefore,

Objective function = Minimize [0.4 × (error/Max. error) + 0.4 × (weight/Max. weight) + 0.2 × (efficiency/Max. efficiency)]

4.2.1.2 Decision Variables

Number of teeth on pinion (z)
Pitch circle diameter of pinion (pcd)

4.2.1.3 Constraints

Induced bending stress (σ_b) ≤ Allowable bending stress of the material
Induced crushing stress (σ_c) ≤ Allowable crushing stress
Induced shear stress (τ_s) ≤ Allowable shear stress
Center distance (d) ≤ Minimum center distance
Pitch line velocity (V) ≤ Maximum pitch line velocity

4.2.1.4   Determination of Range of Pitch Circle Diameter for Pinion

$$a_{max} = \frac{b_{max}}{b/a}$$

$$a_{max} = \frac{d_{1max} + i(d_{1max})}{2}$$

$$d_{1max} = \frac{2a_{max}}{i+1}$$

$$a_{min} = (i+1)\sqrt[3]{\left(\frac{0.74}{\sigma_c}\right)^2 \times \left(\frac{EM_t}{i\lambda}\right)}$$

$$M_t = \frac{60P}{2\pi N}$$

$$D_{1min} = \frac{2a_{min}}{i+1}$$

where
    a = centre distance between two gears (mm)
    b = face width (mm)
    d = diameter (mm)
    E = Young's modulus
    λ = ratio of face width to module
    i = gear ratio
    σ_c = crushing stress (N/mm²)
    P = power (kW)
    N = speed (rpm)
    m = module (mm)
    M_t = torque (N-mm)

The maximum allowable face width (b_{max}) and b/a ratio is taken from the database.

4.2.1.5   Determination of Range of Teeth for Pinion

$$M_{max} = \frac{d_{max}}{Z_{min}}$$

$$M_{min} = \frac{d_{min}}{Z_{max}}$$

Z_{min} is selected such that no interference exists between the mating gears. Taking $M_{min} = 1$ mm, we have $Z_{max} = d_{min}$. The above yields

$$d_{min} \leq d \leq d_{max}$$

$$Z_{min} \leq Z \leq Z_{max}$$

4.2.1.6 Stress Constraints

$$\text{Induced } \sigma_c = 0.74 \left(\frac{i+1}{a} \right) \sqrt{\frac{i \times 1}{i \times b} \times E[M_t]} \leq [\sigma_c] \text{ Allowable}$$

$$\text{Induced } \sigma_b = \frac{i+1}{a.m.b.y} \ [M_t] \leq [\sigma_b] \text{ Allowable}$$

$$\text{Induced } \iota_p = \frac{[M_t] \times 16}{\pi d^3} \leq [\iota_p] \text{ Allowable}$$

4.2.1.7 Efficiency of Coplanar Gears

In applications where large amounts of power are transmitted, the efficiency of gears becomes very important. This requires the most efficient gear designs that conserve energy as well as reduce the amount of heat rejected to the geared system. Generally nonparallel, nonintersecting axis drives have higher sliding or rubbing velocities, which results in higher losses than parallel axis and intersecting axis drives. Consequently, the latter types are most efficient. For this reason, spur, helical, or bevel gears usually are desired for higher power transmission applications unless the design conditions are such that other types of gear drives have more attractive features.

The overall efficiency of spur gears, or all gears for that matter, depends on three separate and distinct types of losses. These three types are commonly known as (1) windage and churning losses, (2) bearing losses, and (3) gear-mesh losses. The efficiency calculation based on gear-mesh follows.

4.2.1.8 Calculation of Efficiency and Weight

$$\eta = 100 - P_1$$

$$P_1 = \frac{50 \times f}{\cos \phi} \left(\frac{H_s^2 + H_t^2}{H_s + H_t} \right)$$

$$|H_s| = (i+1) \sqrt{\left(\frac{R_A}{R} \right) - \cos^2 \phi - \sin \phi}$$

$$H_t = \left(\frac{i+1}{i} \right) \sqrt{\left(\frac{r_A}{r} \right)^2 - \cos^2 \phi - \sin \phi}$$

$$Weight = \frac{\pi}{4} d^2 \times b \times \rho$$

where

P_1 = percent power loss

f = average coefficient of friction

ϕ = pressue angle

H_s = specific sliding velocity at start of approach action (m/min)

H_t = specific sliding velocity at end of recess action (m/min)

i = gear ratio

r_A = addendum circle radius of pinion (mm)

r = pitch circle radius of pinion (mm)

R_A = addendum circle radius of gear (mm)

R = pitch circle radius of gear (mm)

4.2.1.9 Error

An error is the value obtained by subtracting the design values of a dimension from its actual value. Errors can be broadly classified into two main categories: individual errors and composite errors.

4.2.1.10 Individual Errors

Individual errors involve those errors that are deviations of individual parameters of toothing from their ideal values. Under such headings fall profile error, adjacent pitch error, base pitch error, tooth alignment error, radial run-out error, tooth thickness error and so on. These errors are measured by special measuring instruments and the different types of individual errors are discussed below.

4.2.1.11 Profile Error

Profile error is an indication of departure of the actual profile from the ideal involute profile. This departure at any point is measured normal to the involute profile.

4.2.1.12 Pitch Error

Pitch error denotes the departure of the actual spacing of the teeth from the ideal one. The adjacent pitch error is the departure measured on similar flanks of two edges of teeth. When the measurement is done over a length more then one pitch apart, namely k number of pitches, it is called a cumulative pitch error. Base pitch error is the difference between the actual and ideal base pitch.

4.2.1.13 Tooth Alignment Error

Tooth alignment error is also known as the error of distortion. When a spur gear is cut, its tooth traces should follow the ideal path, i.e., parallel to the axis of the gear. The tooth alignment error is an indication of deviation from the ideal path. This error is usually measured in micrometers over a given distance on the tooth width of the gear.

4.2.1.14 Radial Run-Out Error

Radial run-out error is a measure of the eccentricity of the tooth system.

4.2.1.15 Axial Run-Out Error

Axial run-out error is a measure of the out of true position of the axis of the gear and is measured by placing a dial gauge over a reference surface whose axis is held at a specific distance and parallel to the axis of rotation of the gear.

4.2.1.16 Tooth Thickness Error

Tooth thickness error is the value obtained by subtracting the tooth design thickness from the actual tooth thickness measured along the surface of the reference of the pitch cylinder.

4.2.1.17 Base Circle Error

Base circle error denotes the difference between the actual and the theoretical dimension of the base circle diameter.

4.2.1.18 Composite Error

Composite error is the combined effect of all the individual errors. These errors include profile error, pitch error, tooth alignment error, radial run-out error, axial run-out error, and so on. Because measuring each and every individual error of each and every product is not practical in a regular production schedule, the composite error is used. Thus,

Total error = profile error + pitch error + tooth alignment error + radial run-out error + axial run-out error

4.2.2 Applying Genetic Algorithm

The flow chart of the optimization procedure based on GA is given in Figure 4.1. These steps consist of the following items.

4.2.2.1 Coding

The decision variables, namely pitch circle diameter (PCD) and the number of teeth in pinion gear (Z) are coded as binary. Because the accuracy of Z cannot be fractional, it is coded for a whole number. PCD coding can be done as fractions. Initially, a population is created using random numbers.

4.2.2.2 Gene

Each bit of the binary number (either 0 or 1) is treated as gene in order to apply GA. The complete binary number is treated as chromosome.

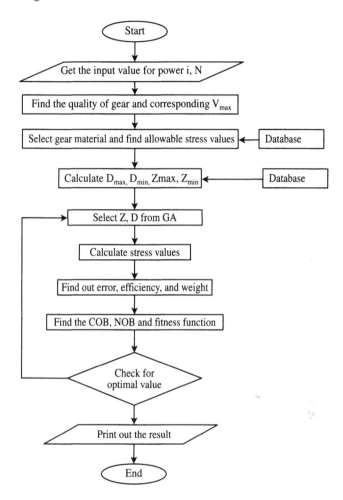

FIGURE 4.1 Flow chart representing the GA process.

4.2.2.3 Chromosome Length

Chromosome length is a major parameter. Due to multivariable objective functions, each chromosome (also called a string) will contain substrings. The number of substrings will depend upon the number of variables in the objective function. Substring lengths will depend on the range of PCD and Z. The length is calculated as explained in the following illustration.

4.2.2.4 Crossover

The string length employed in this work is large. The larger the range of PCD and Z, the larger the string length will be. Simple single-point crossover will reduce the efficiency of the algorithm. Therefore, multipoint crossover is employed to achieve better results. Two random numbers are generated, one each for PCD and Z. If the two numbers have the probability to undergo crossover, it is then performed separately between the strings to the right of the two selected sites.

4.2.2.5 Mutation

Mutation is a bit-by-bit process. A random number is generated for each and every bit of all the chromosomes. If it satisfies the mutation probability, the bits are swapped from 0 to 1 or vice versa. This operator is employed with the intention to better the solution for a faster convergence to the optimal value.

4.2.2.6 Numerical Illustration

Design a gear-set for an IC engine to transmit a power of 1.4 kW. The speed of the crankshaft gear (pinion) is 3000 rpm and the gear ratio is 2.

4.2.2.7 Initialization

Variable range: $18 \leq Z \leq 38$ and $32.103 \leq d \leq 333.33$
The coding scheme is given in Table 4.1.
Chromosome length = 5 + 12 = 17.

Since a few decoded values of Z can go out of range, this condition is taken as a constraint violation and combined with the objective function. Therefore,

NOF = Minimize [COF + stress constraint violation + teeth constraint violation]

4.2.2.8 Evaluation

The decoded values from a string vary from 0 to 2^n. However, the actual values may not vary in that range. The decoded values of the strings should be mapped in the range in which the actual values will vary. Once all the values are decoded, they should be mapped as given below:

$$\text{Mapped value} = \text{lower limit} + [(\text{range} \times \text{decoded value})/2^n - 1]$$

where $(\text{range}/2^n - 1)$ is called the precision of the variable.

TABLE 4.1
Initialization — Variables Coding

Coding for 'Z'	Coding for 'd'
Range of $Z = Z_{max} - Z_{min} = 19$	Range of $d = d_{max} - d_{min} = 301.297$
2^n = Range/Accuracy	2^n = Range/Accuracy
Accuracy for $Z = 1$, because Z cannot be a decimal value	Accuracy for $d = 0.1$
$n = 5$	$n = 12$
Possible combinations = $2^5 = 32$	Possible combinations = $2^{12} = 4096$

TABLE 4.2
Initial Random Population

S#	Z	PCD	ZDV	PCD_{DV} (mm)	Error (μm)	Weight (kg)	η%	Objective Function
7	11101	110110100011	477	282.00	332.23	24.20	99.27	100.36
2	01011	111100011001	29	200.00	377.09	17.00	98.82	0.413
*3	00101	001000001111	23	69.00	240.44	0.88	98.55	0.016
4	01100	111001010111	30	300.30	400.79	55.5	98.65	0.454
5	01100	011101100111	30	165.00	299.30	9.23	98.85	0.145
6	00000	011100001011	18	162.00	336.18	14.56	98.21	0.713

* Best sample

After all the variables are mapped, they are used to calculate the objective function value. A sample output for the initial population by computer simulation is given below.

4.2.2.8.1 Initial Population

The initial random population is shown in Table 4.2.

4.2.2.8.2 Population Obtained after First Generation

The population obtained after first generation is shown in Table 4.3.

After applying the GA operators, a new set of population is created. The values are then decoded and objective function values are calculated. This completes one generation of GA. Such iterations are continued until the termination criteria are achieved. The above process is simulated by computer program with a population size of 20 iterated for 50 generations.

TABLE 4.3
Population Obtained after First Generation

Initial Population		After Reproduction		After Crossover		After Mutation	
Z	PCD	Z	PCD	Z	PCD	Z	PCD
11101	110110100011	01100	111001010111	01101	111001010111	01101	111001110111
01011	111100011001	01011	001000001111	00100	001000001111	11100	011000001111
00101	001000001111	01011	111100011001	01011	101000001111	00011	010001001111
01100	111001010111	00101	001000001111	00101	011100011001	01100	011100011001
01100	011101100111	01100	011101100111	01100	011101101011	01100	000011101011
00000	011100001011	00000	011100001011	00000	011100000111	00010	011100010110

TABLE 4.4
Evaluation of First Generation

S#	Z	PCD	ZDV	PCD$_{DV}$ (mm)	Error (μm)	Weight (kg)	η%	Objective Function
1	01101	111001110111	31	279.00	364.01	43.19	98.88	0.380
2	11100	011000001111	46	138.00	264.52	3.522	98.22	90.08
3	00011	010001001111	21	252.00	390.61	46.98	98.43	0.351
4	01100	011100011001	30	165.00	299.20	9.23	98.85	0.245
*5	01100	000011101011	30	52.50	219.51	0.297	98.85	0.009
6	00010	011100010110	20	160.00	324.00	12.62	98.36	0.161

* Best sample

4.2.2.8.3 Evaluation of First Generation

After the first generation, all the samples are evaluated and shown in Table 4.4.

From the above data, the best point in the population is improved after every generation. Table 4.5 compares the values obtained through conventional problems and applies GA for the above mentioned illustration.

4.2.3 APPLYING SIMULATED ANNEALING ALGORITHM

The simulated annealing procedure simulates the process of slow cooling of molten metal to achieve the minimum function value. Controlling a temperature-like parameter introduced with the concept of the Boltzmann probability distribution simulates the cooling phenomenon.

The flow chart for this process is given in Figure 4.2. The following steps explain the procedure of the simulated annealing algorithm.

Step 1: Here an initial value of teeth (Z) and pitch circle diameter for pinion (d) is chosen. The value is taken as the mean of the extreme value of the range. The temperature parameter T is taken as the average value of the function at four extreme points, which should be sufficiently high. The number of iterations to be performed at a particular temperature is taken as 20. The termination criteria chosen for this problem is when the temperature falls below 0.001°C. The temperature will be reduced by a factor of 0.9 after each iteration to ensure slow cooling.

TABLE 4.5
Optimization Results

	Z	PCD (mm)	Error (μm)	η%	Weight(kg)
Conventional method	21	38.1	212.43	98.24	0.162
Genetic algorithm	28	35	204.32	98.77	0.090

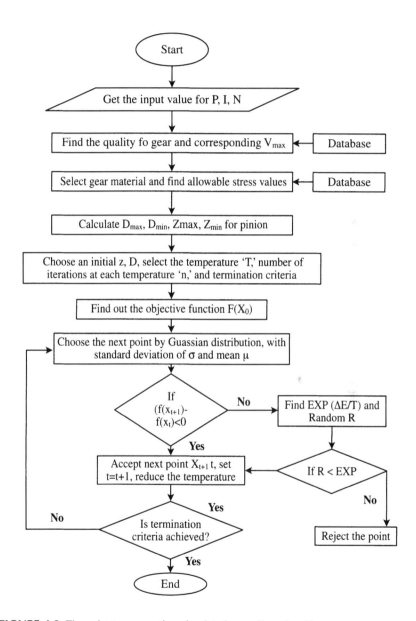

FIGURE 4.2 Flow chart representing simulated annealing algorithm.

Step 2: A neighboring point is created by a normal distribution as

$$X_x = x_1 + \left(\sum_{i=1}^{n} r_i - \frac{n}{2} \right)$$

where

σ = standard deviation

n = number of random numbers to be generated.

Step 3: Find the new function value and calculate the difference between the two function values $\Delta E = f(x_2)\ f(x_1)$. If this ΔE is negative, the latter point created is a better point and hence it is immediately accepted; if ΔE is positive, accept that point based on the Boltzmann probability given as

$$P(E(t + 1)) = \min[1,\ e^{E/KT}]$$

where

$P(E(t + 1))$ = probability of acceptance of a point

T = temperature parameter

K = Stefan–Boltzmann constant that depends on the type of problem. For positive values of ΔE, a random number between 0 and 1 is created. If this random number is less than $e^{\Delta E/KT}$, the latter point is accepted, or else it is rejected. From the above equation, at a higher temperature every point has an equal probability of acceptance and at lower temperatures the probability of acceptance of worse points decreases.

Step 4: Check whether the termination criteria are achieved for T, or else go to Step 2.

In this problem, the Stefan–Boltzmann constant K is taken as 0.05 because the best performance is achieved for this value. The initial temperature is found to be 500°C. A computer program simulates the above process, and the results obtained by the simulated annealing algorithm are given in Table 4.6.

At the initial stages, the variation of the objective function is very high. As the temperature parameter reduces, the variation is also reduced, convergence is achieved; and in the final stage, the objective function converges to a minimum value irrespective of the variation of other parameters like error, efficiency, and weight.

4.2.3.1 Gear Details (without Optimization)

Number of teeth (Z_1) = 21
Tooth depth = 0.1406 = 3.57 mm

TABLE 4.6
Result Obtained by Simulated Annealing Algorithm

	Z	PCD (mm)	Error (μm)	η%	Weight (kg)
Simulated Annealing	28	35	204.32	98.77	0.090

Pitch circle diameter = 1.5 = 38.1 mm
Addendum diameter = 1.625 = 41.275 mm
Span over three teeth = 0.545
Material = cast iron grade 17
Maximum temperature = 270°
Backlash with mating gear = 0.006, 0.008
Hardness = 200–240 BHN
Face width (b) = 0.430/0/420
Diameter pitch = 0.55/mm

4.2.3.2 Details of the Optimized Gear

Number of teeth (Z_1) = 28
Pitch circle diameter (PCD) = 1.378 = 35mm
Diameter pitch = 0.8/mm
Material = cast iron grade 17
Face width = 0.492 = 12.5 mm
Tooth depth = 0.11073 = 2.1825 mm
Module = 1.25 mm
Addendum diameter = 1.5 = 38.125 mm
Span over three teeth = 0.369 = 9.375 mm

The data for quality of gears (spur) is given in Table 4.7. The data for material selection is given in Table 4.8 and the data for errors is given in Table 4.9.

4.3 DESIGN OPTIMIZATION OF THREE-BAR TRUSS

4.3.1 PROBLEM DESCRIPTION

The three-bar truss shown in Figure 4.3 is subjected to a load P inclined at an angle to the horizontal at node 4, as shown. Due to this load P, stress and deflections are produced in the members 1, 2, and 3 of the truss. In this problem, the nodes are separated at a distance 1 m apart in the horizontal and vertical directions. The main objective of the problem is to find the optimal values of the cross-sectional area of the members 1, 2, and 3 to withstand the given load P and also the distance between the two nodes 1 and 4. In this problem, we are assuming the cross-sectional area of members 1 and 3 are the same and the cross-sectional area of member 2 is different. Several cross sectional shapes are available for truss members but a hollow circular cross-section is chosen for members 1, 2, and 3.

4.3.2 DESIGN VARIABLES

In this problem, the design variables are the cross-sectional area of members 1 and 3, the cross-sectional area of member 2, and the member length, l.

A_1 = Cross-sectional area of members 1 and 3 (m²)
A_2 = Cross-sectional area of member 2 (m²)
l = Length between the nodes (m)

TABLE 4.7
Database for Quality of Gears (Spur)

Category	Types of Equipment/Application	Quality of Gears (IS Code)	Maximum Allowable Velocity (m/s)	Machining Process
General mechanical equipment and machine tools	1. Host and conveying systems	7 to 12	0.8	Cast, press forged
	2. Internal combustion engines	5 to 9	8	Fine generated
	3. Steam engines	6 to 11	5	Milled generated
	4. Turbines	5 to 7	8	Fine finished
	5. Textile industries	6 to 12	1.2	Machined with form cutter
	6. Chemical industries	6 to 10	5	Milled generated
	7. Printing industries	6 to 9	6	Milled generated
	8. Railway and signaling equipment	7 to 12	1.2	Machined with form cutter
	9. General purpose machine tools	5 to 9	8	Milled generated
Fine mechanics	1. Watch industries	5 to 9	15	Fine finished
	2. Calculation and business machines	8 to 12	0.8	Cast, press forged
Transport machines	1. Aircraft industries	3 to 8	15	Finished by grinding, scraping etc.
	2. Cars and trucks	5 to 8	8	Fine finished
	3. Locomotives and similar machines	6 to 12	1.2	Machined with form cutter
	4. Agricultural equipment, tractors, etc.	7 to 12	1.2	Cast, press forged
	5. Marine engines	5 to 8	5	Milled generated
Testing machines	1. Measuring instruments	2 to 5	15	Finished by grinding, scraping, etc.
	2. Master gears	1 to 3	15	Finished by grinding

TABLE 4.8
Database for Material Selection

Category	Power (P, kW) and Material	Young's Modulus (E)	Crushing Stress (σ_c, N/mm²)	Bending Stress (σ_b, N/mm²)	Density (ρ) (kg/m³)
General purpose machine tools	If P < 2, C1 Grade 70	1.1×10^5	375	126	7850
	If 2 < P < 4, C45 steel	2.15×10^5	500	135	7850
	If 4 < P < 10, C55 Mn75	2.15×10^5	900	542	7850
	If 10 < P < 15, 15Ni2Cr1Mo15	2.15×10^5	950	320	7850
	If 15 < P < 20, AISI 2340 steel	2.03×10^5	988	740	7814
	If P > 20, 40Ni2Cr1Mo28	2.15×10^5	1100	400	7857
Fine machines	If P < 0.05, polysterene	2.8×10^5	31.56	24	1050
	If 0.05 < p < 0.1, cast phenolic	4.9×10^5	35	27	1325
	If P > 0.1, Nylatron Gs	4.2×10^5	42	32	1132
Transport machines	If Aircraft, 75 S-T alum plate	7.29×10^5	288	215	2761
	If Marine, titanium alloy	1.2×10^5	526	392	4500
	If P < 20, AISI 2340 steel	2.03×10^5	988	740	7814
	If 20 < P < 50, 15Ni2Cr1Mo28	2.15×10^5	950	320	7850
	If P > 50, 40Ni2Cr1Mo28	2.15×10^5	1100	400	7857

TABLE 4.8 (Continued)
Database for Material Selection

Category	Power (P, kW) and Material	Young's Modulus (E)	Crushing Stress (σ_c, N/mm²)	Bending Stress (σ_b, N/mm²)	Density (ρ) (kg/m³)
Testing machines	If P < 0.1, steel iolite	3×10^5	123	92	7013
Machine tools	If P < 2, C1 Grade 70	1.1×10^5	375	126	7850
	If 2 < P < 4, C45 steel	2.15×10^5	500	135	7850
	If 4 < P < 10, C55 Mn 75	2.15×10^5	900	542	7850
	If 10 < P < 15, 15Ni2Cr1Mo15	2.15×10^5	950	320	7850
	If 15 < P < 20, AISI 2340 steel	2.03×10^5	988	740	7814
	If P > 20, 0Ni2Cr1Mo28	2.15×10^5	1100	400	7857

4.3.3 OBJECTIVE FUNCTION

The main objective of this problem is to minimize the weight or volume of the entire structure so that the structure can withstand the given applied load, P. If the material or volume of the structure is reduced, this automatically reduces the cost of the truss.

TABLE 4.9
Database for Errors

Gear Quality	Total Error
3	$30.1 + 2 m + 0.5\sqrt{b} + 1.265\sqrt{d}$
4	$45.7 + 3.1 m + 0.63\sqrt{b} + 1.9908\sqrt{d}$
5	$70 + 4.76 m + 0.80\sqrt{b} + 3.135\sqrt{d}$
6	$105.3 + 7.6 m + \sqrt{b} + 4.938\sqrt{d}$
7	$145 + 10.08 m + 1.25\sqrt{b} + 6.99\sqrt{d}$
8	$191 + 14.25 m + 2\sqrt{b} + 9.566\sqrt{d}$
9	$252 + 18.64 m + 3.15\sqrt{b} + 11.1526\sqrt{d}$
10	$335 + 24.6 m + 5\sqrt{b} + 18.083\sqrt{d}$
11	$443 + 32.6 m + 8\sqrt{b} + 24.75\sqrt{d}$
12	$600 + 43.5 m + 12.5\sqrt{b} + 34.41\sqrt{d}$

m = module, b = face width, and d = diameter

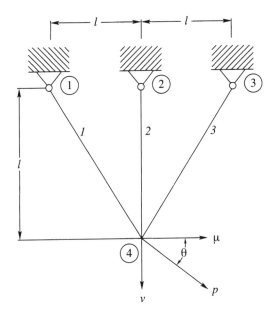

FIGURE 4.3 Three-bar truss.

Volume of a member = cross-sectional area × length

Total volume of the structure = volume of member 1 + volume of member 2 + volume of member 3

$$V = \sqrt{2}A_1 l + A_2 l + \sqrt{2}A_1 l$$

$$V = l\left(\sqrt{2}A_1 + A_2 + \sqrt{2}A_1\right)$$

$$V = l\left(2\sqrt{2}A_1 + A_2\right)$$

The objective is to minimize this last function for V. In order to include the constraints given below, this is converted to an equivalent unconstrained problem by adding penalty terms to the above objective function according to each constraint violation.

4.3.4 Design Constraints

Nine design constraints must be satisfied to give the optimal value of cross-sectional areas and length. These constraints are stress constraints, deflection constraints, frequency constraints, buckling constraints, and area constraints.

4.3.5 STRESS CONSTRAINTS

This constraint gives the value of the stress in various members of the truss and these values should be less than the maximum allowable stress value.

Let σ be the stress in members 1, 2, and 3 under the load P. These values are calculated from the forces in the members as follows:

$$\sigma_1 = \frac{1}{\sqrt{2}\left(\frac{P_u}{A_1} + \frac{P_v}{A_1 + \sqrt{2}A_2}\right)}$$

$$\sigma_2 = \frac{\sqrt{2}P_v}{A_1 + \sqrt{2}A_2}$$

$$\sigma_3 = \frac{1}{\sqrt{2}\left(\frac{P_v}{A_1 + \sqrt{2}A_2} - \frac{P_u}{A_1}\right)}$$

From the equation above, the value of σ_1 is always larger than σ_3. Therefore, constraints need to be imposed on only σ_1 and σ_2. If σ_a is the allowable stress for the material, the stress constraints are $\sigma_1 \le \sigma_a$ and $\sigma_2 \le \sigma_a$.

$$\frac{1}{\sqrt{2}\left(\frac{P_u}{A_1} + \frac{P_v}{A_1 + \sqrt{2}A_2}\right)} \le \sigma_a$$

$$\frac{\sqrt{2}P_v}{A_1 + \sqrt{2}A_2} \le \sigma_a$$

4.3.6 DEFLECTION CONSTRAINTS

Using analysis procedures for statically indeterminate structures, the horizontal and vertical displacements u and v of node 4 of the truss are

$$u = \frac{\sqrt{2}lP_u}{A_1 E}$$

$$v = \frac{\sqrt{2}lP_v}{\left(A_1 + \sqrt{2}A_2\right)E}$$

where E is the Young's modulus of the material and P_u and P_v are the horizontal and vertical components of the load P. These horizontal and vertical deflections of node 4 must be within the specified limits u and v, respectively,

$$\frac{\sqrt{2}lP_u}{A_1E} \leq \Delta u$$

$$\frac{\sqrt{2}lP_v}{\left(A_1 + \sqrt{2}A_2\right)E} \leq \Delta v$$

4.3.7 FREQUENCY CONSTRAINTS

The fundamental natural frequency of the structure should be higher than the specified frequency. When the natural frequency of the system matches the external frequency, resonance will occur. At that time, the amplitude of vibration is a maximum and will cause the failure of the structure. The lowest (fundamental) natural frequency of the structure must be as high as possible to avoid any possibility of resonance. This constraint also makes the structure stiffer. Frequencies of a structure are obtained by solving an eigenvalue problem involving stiffness and mass properties of the structure. The lowest eigenvalue related to the lowest natural frequency of the symmetric three-bar truss is completed using a consistent mass model as

$$\xi = \frac{3EA_1}{\rho l2\left(4A_1 + \sqrt{2}A_2\right)}$$

where ρ is the mass density (kg/m^3). This value should be greater than $(2\Pi_0)^2$.

4.3.8 BUCKLING CONSTRAINTS

Buckling constraints are expressed as

$$-F_i \leq \frac{\Pi^2Ei}{li^2}$$

where $I = \{1, 2, 3\}$. The negative sign for F_i is used to make the left-hand side of the constraints positive when the member is in compression. No need exists to impose buckling constraints for members under compression; the dependence of the moment of inertia I on the cross-sectional area of the members must be specified. The most general form is

$$I = B \cdot A^2$$

where A is the cross-sectional area and is a nondimensional constant. This relation follows if the shape of the cross-section is fixed and all its dimensions are varied in the same proportion. The axial force for the ith member is given as

$$F_i = A_i$$

where $i = \{1, 2, 3\}$ with tensile force taken as positive. Members of the truss are considered columns with pin ends. Therefore, the buckling load for the ith member is given as

$$\frac{\Pi^2 Ei}{li^2}$$

where l is the length of the ith member.

$$-\frac{1}{\sqrt{2}\left(\frac{P_u}{A_1} + \frac{P_v}{A_1 + \sqrt{2}A_2}\right)} \le \frac{\Pi^2 E\beta A_1}{2l^2}$$

$$-\frac{\sqrt{2}P_v}{A_1 + \sqrt{2}A_2} \le \frac{\Pi^2 E\beta A_2}{l^2}$$

$$-\frac{1}{\sqrt{2}\left(\frac{P_v}{A_1 + \sqrt{2}A_2} - \frac{P_u}{A_1}\right)} \le \frac{\Pi^2 E\beta A_1}{2l^2}$$

4.3.9 AREA CONSTRAINT

The cross-sectional areas A_1 and A_2 must both be nonnegative, i.e., $A_1, A_2 \ge 0$. Most practical problems would require each member to have a certain minimum area, A_{min}. The minimum area constraints can be written as

$$A_1, A_2 \ge A_{min}$$

The optimum design problem is to find the cross-sectional areas $A_1, A_2 \ge A_{min}$ and the length to minimize the volume of the structure subjected to the constraints. This problem consists of three design variables and nine inequality constraints. To handle the constraints, penalty terms are added to the objective function according to the degree of each constraint violation.

4.4 SPRING DESIGN OPTIMIZATION

A coil spring (Figure 4.4) is used for storing energy in the form of resilience. Coil springs are mainly used with independent suspension in vehicles. In coil springs, the energy stored per unit volume is almost double that of leaf springs. A coil spring does not have noise problems nor does it have static friction, causing harshness of ride like leaf springs. The spring takes the shear as well as bending stresses.

FIGURE 4.4 Coil spring.

4.4.1 PROBLEM FORMULATION

Consider a coil spring that is subjected to an axial load P as shown in Figure 4.4. Due to this load P, the wire experiences twisting and some elongation of the spring takes place. The problem is to design a minimum-mass spring that carries given loads without material failure, while it satisfies other performance requirements.

4.4.2 DESIGN VARIABLES

In this problem, three design variables are chosen: wire diameter d, mean coil diameter D, and the number of active coils, N.

4.4.3 OBJECTIVE FUNCTION

The main objective of this problem is to minimize the mass of the spring in order to carry given loads without material failure.

$$M = \pi(N + Q) \, \Pi^2 D d^2$$

N = Number of active coils
Q = Number of inactive coils
D = Coil diameter (m)
d = Wire diameter (m)
ρ = mass density of material ($N - S^2/m^4$)

To include the constraints given below, this is converted to an equivalent unconstrained problem by adding penalty terms to the above objective function according to each constraint violation.

4.4.4 DESIGN CONSTRAINTS

Four design constraints must be satisfied to give the optimal values of d, D, and N. These constraints are deflection constraint, shear stress constraint, constraint on frequency of surge waves, and diameter constraint.

4.4.5 Deflection Constraint

The deflection under the given load P is often required to be at least. Therefore, calculated deflection must be greater than or equal to $\Delta \times 2$. This form of the constraint is common to spring design.

$$\frac{8PD^3N}{Gd^4} \geq \Delta$$

4.4.6 Shear Stress Constraint

To prevent material over-stressing, shear stress in the wire must be less than τ_a.

$$\frac{8PD}{\Pi d^3 \left(\frac{4D-d}{4D-4d} + \frac{0.015d}{D} \right)} \leq \tau_a$$

4.4.7 Constraint on Frequency of Surge Waves

When a load acting on the spring is removed, the spring executes a vibratory motion. During this vibratory motion, a chance of resonance exists. To avoid this case, the frequency of surge waves along the spring must be as large as possible.

$$\frac{d}{2\Pi D^2 N} \sqrt{\frac{G}{2\rho}} \geq \omega_0$$

4.4.8 Diameter Constraint

The outer diameter of the spring should not be greater than $\bar{D}$,

$$D + d \cdot \bar{D}$$

4.4.9 Limits on Design Variables

To avoid practical difficulties, minimum and maximum size limits put on the wire diameter, coil diameter, and number of turns are

$$d_{min} \leq d \leq d_{max}$$

$$D_{min} \leq D \leq D_{max}$$

$$N_{min} \leq N \leq N_{max}$$

4.4.10 Implementation of Genetic Algorithm

4.4.10.1 Three-Bar Truss

4.4.10.1.1 Input Data

Allowable stress:	Members 1 and 3, $\sigma 1a = \sigma 3a = 5000$ psi
	Member 2, $\sigma 2a = 20,000$ psi
Allowable displacements:	Ua = 0.005 in
	Va = 0.005 in
Modulus of elasticity:	E = 1 × 107 psi
Weight density:	$\gamma = 1 \times 101$ lb/in3
Constant:	$\beta = 1.00$
Lower limit on frequency:	2500 Hz

The upper and lower limits of the design variables are selected as follows:

$$1.5 \leq x_1 \leq 2.0 \text{ (in)}$$
$$1.5 \leq x_2 \leq 2.0 \text{ (in)}$$
$$80 \leq x_3 \leq 100 \text{ (in)}$$

4.4.10.2 Spring

4.4.10.2.1 Input Data

Number of inactive coils:	$Q = 2$
Applied load:	$P = 44.4$ N
Shear modulus:	$G = 7921.2$ E +.7 N/m^2
Minimum spring deflection:	$\Delta = 0.0127$ m
Weight density of spring material:	$\gamma = 77,319.9$ N/m^3
Gravitational constant:	$g = 9.81$ m/s^2
Mass density of material:	$\rho = 7889.7$ N $-$ S^2/m^4
Allowable shear stress:	$\tau = 55,104$ E + 4 N/m^3
Lower limit on surge wave frequency:	$\omega = 100$ Hz
Limit on the outer diameter:	$\bar{D} = 0.038$ m

The upper and lower limits of the design variables are selected as follows:

$$0.05 \times 10^2 \leq d \leq 0.2 \times 10^2 \text{ (m)}$$
$$0.25 \times 10^2 \leq D \leq 1.3 \times 10^2 \text{ (m)}$$
$$2 \leq N \leq 15$$

4.4.11 GA Parameters

Number of digits: 10 (for variables x_1, x_2, x_3, d, D)
4 (for variable N)

Three bar truss:	Total 30 digits
Coil spring:	Total 24 digits

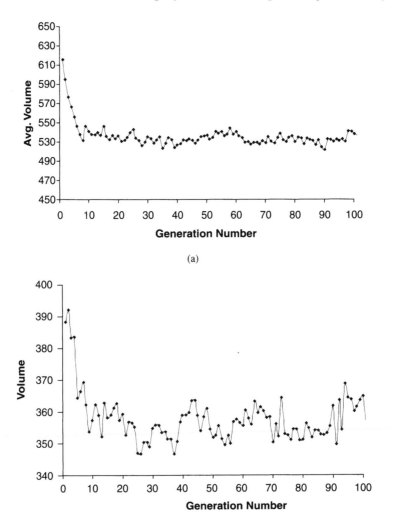

FIGURE 4.5 (a) Optimization results of three-bar truss design (Average volume). (b) Optimization results of three-bar truss design (Volume).

Sample size: 30
Selection operator: Tournament selection
Crossover probability (P_c): 0.7
Mutation probability (P_m):   0.01
Number of generations: 100

Results obtained by GA are shown in Figure 4.5 for the three-bar truss and Figure 4.6 for the spring.

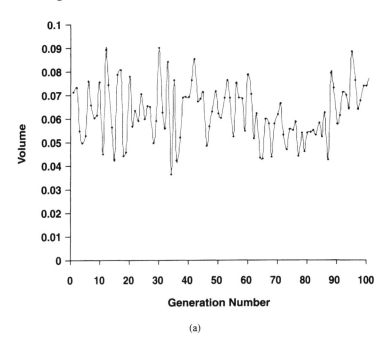

(a)

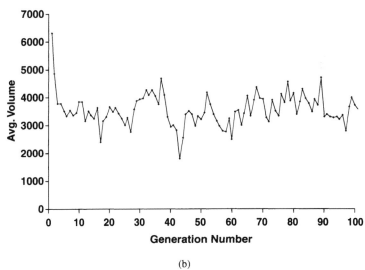

(b)

FIGURE 4.6 (a) Optimization results of spring design (Volume). (b) Optimization results of spring design (Average volume).

4.5 DESIGN OPTIMIZATION OF SINGLE-POINT CUTTING TOOLS

The basic elements of the modern metal removal process consist of a machine tool, a control system, and the cutting tool. The common method of metal removal is using an edged cutting tool. Thus, a control system and machine tool are useless without the cutting tool and vice versa. Most machining operations can be performed by the application of standard cutters. To select the proper tool effectively, a good knowledge of the metal cutting process and familiarity with tool geometry is necessary. Standard cutting tools should be used whenever possible for reasons of economy. Cutting tool manufacturers mass-produce their products and are able to keep the cost to a minimum. In this technological era, the production of tools and their marketing has become very competitive. For minimizing the cost of a tool, the tool life should be increased so that the tool can be used for more jobs; the overall cost will then be minimized. While trying to increase the lifetime, the accuracy and effectiveness of the working of the tool should also be kept in mind and the machinability should not be reduced.

4.5.1 SINGLE-POINT CUTTING TOOLS

Single-point cutting tools (SPCT) have served to demonstrate metal cutting principles and other types of cutting tools are cited at appropriate places. These tools use various methods to work metals. In general, these methods can be divided into two classes, cutting tools and forming tools. Cutting tools remove metal in the form of chips, whereas forming tools either deform metals by making them flow into new shapes or by shearing them into new shapes. Cutting tools that remove metal in the form of chips may be divided into five basic groups according to machining operations, namely, turning, drilling, milling, shaping (or planning) and grinding tools.

Chip formation is the same in all the machining operations but the method of holding the work is different. The cutting tool materials commonly used are HSS (high speed steel), cast alloy, cemented carbide and cemented oxide. The tool tip is held in the proper position mechanically or permanently soldered or brazed in position.

4.5.2 DEVELOPMENT OF MODEL

4.5.2.1 Overview of Tool Geometry

Consider a single point cutting tool, shown in Figure 4.7 and Figure 4.8. The main elements of a SPCT are

> *Shank*: The main body of the tool.
> *Flank*: The surfaces below and adjacent to the cutting edge.
> *Face*: The surface on which the chip impinges.
> *Nose*: The intersection of the side cutting edge and the end cutting edge.

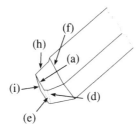

FIGURE 4.7 An isometric view of a cutting tool.

Cutting edges: The portions of the face edges that separate the chip from the workpiece. The main cutting edges are the side cutting edge, the nose, and the end cutting edge.

Cutting-edge angles: The angles between the sides of the tool shank and the corresponding cutting edges. The two main cutting-edge angles are the side cutting-edge and end cutting-edge angle.

Relief angle: The angles between the portion of the side flank immediately below the side cutting edge and a line perpendicular to the base of the tool, measured at right angles to the side blank.

Back relief angle: The angle between the face of the tool and a line parallel with the base of the tool, measured in a perpendicular plane through the side cutting edge.

Side-rake angle: The angle between the face of the tool and a line parallel with the base of the tool, measured in a plane perpendicular to the base and the side cutting edge. The rake angles, the cutting-edge angles, and the nose radius have the greatest influence on the cutting process. Along with the above dimensions of the tool, the machining process is also characterized by some variables in the cutting process itself, such as the feed, the depth of cut, and the cutting velocity.

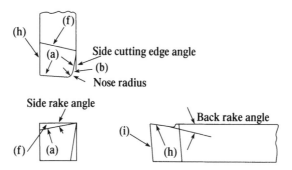

FIGURE 4.8 Three views of cutting tools.

Feed: The rate of cutting tool advance relative to cutting speed. This value is usually expressed in terms of the distance that the tool is fed for each "cycle" of the work.

Depth of cut: The change in the dimension of the material measured in the plane of motion of the cutting tool.

Velocity of cut: The velocity of the workpiece relative to the cutting tool.

4.5.3 DESIGN MODEL

Nomenclature

β_s, β_b	Side, rake and effective rake angle, rad.
β_e	Effective rake angle—equivalent to β in orthogonal cutting, rad.
β	Rake angle for orthogonal cutting, rad.
r_n	Nose radius, in.
ϕ	Friction angle, rad.
δ	Side cutting edge angle, rad.
α	Shear angle, rad.
ι	Back rake angle, rad.
θ	Clearance angle, rad.
σ_{str}	Shear strength of the workpiece material, lb /sq. in
σ_{cut}	Stress developed in the workpiece due to the cutting force, lb/sq. in
F_c	Cutting force, lbs.
F_{max}	Maximum cutting force for a particular nose-radius, lbs.
H_s	Heat dissipated during cutting, BTU.
H_e	Total heat generated during cutting, BTU.
c	Specific heat of the workpiece, BTU/ lb. °C.
G_m	Weight of chip produced per minute, lbs/min.
η	Ratio of heat generated during cutting, which is not removed by cutting fluid.
$\mathfrak{z}$	Mechanical equivalence, ft. lb /BTU.
f	Feed, in/rev.
d	Depth of cut, in.
V	Cutting velocity, ft/min.
T	Tool life, min.
ρ	Density of the workpiece material, lb/ in³.
P_{max}	Maximum power available, lb. ft / min.
c_1	Contact length, in.
t_1, t_2	Thickness of uncut and cut chip, in.
r_c	Ratio of thickness of uncut and cut chip (= t_1/t_2).
A_s	Area of cross-section of chip in the shear plane.
T	Rise in temperature near the tool tip, °C.
W_e	Wear land area, in².

BHN	Brinnel hardness number.
n_1, n_2, n_3, C_1	Constants in Taylor's tool life equation.
f_m, θ	Cost of machining, dollars
f_d, γ	Shear strain in the tool at the point of contact
r_{sf}	Nose radius for providing required surface finish in
$d_i, e_i, g_i (i = 0, \ldots 2)$	Coefficients in surface finish expressions.

The mechanics of metal cutting are well known and have been used in several optimization studies. Now the mathematical model for the SPCT is derived. The design model is concerned with obtaining the dimensions of the tool that result in the best performance of the tool, whereas the manufacturer is inclined to choose machining conditions that minimize the cost of machining the tool.

The designer has control over the dimensions of the tool — especially the side and back rack angles, β_s and β_b, respectively, and the nose radius, r_n. The manufacturer has control over condition of cutting, the velocity of cut (v), the feed (f) and depth of cut (d). The possible measures of good performance of the tool is to reduce the shear strain, γ, caused by the tip of the tool during the cuttingprocess. A high shear strain in the workpiece during machining increases the power required for cutting and the temperature at the tip of the tool. A work-hardening effect is also produced due to high strain. The expression of γ is obtained as

$$\gamma = \frac{\cos(\beta_e)}{\sin \alpha \cos(\alpha - \beta_e)}$$

where β_e is the rack angle of the tool in orthogonal cutting. The two types of cutting introduced here are orthogonal cutting and oblique cutting. It is shown in Fig. 4.9.

The analysis for orthogonal cutting is much simpler than oblique cutting, where the direction of chip flow as shown in Figure 4.9 makes an angle ψ that is not equal to zero with the workpiece. However, a single-point cutting tool used

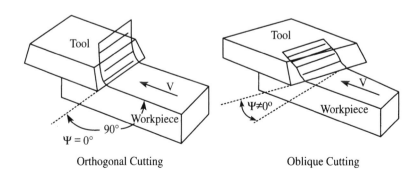

FIGURE 4.9 Types of cutting: (a) orthogonal cutting; (b) oblique cutting.

in turning and with more than one cutting edge is an oblique cutting tool. The effective rack angle β_e plays the same role as a normal rate angle in orthogonal cutting. The angle ψ can be written in terms of β_e, β_s, β_b, and δ as

$$\tan \psi = \tan \beta_b \cos \delta - \sin \delta \tan \beta_s$$

The effective rack angle β_e is defined in terms of β_b and β_s as

$$\tan \beta_c = \tan \beta_b \sin \delta - \cos \delta \tan \beta_s$$

To obtain the shear angle, α,

$$\tan \alpha = \frac{\gamma_c \cos \beta_e}{1 - \gamma_c \sin \beta}$$

where, γ_c is ratio of chip thickness t_2/t_1. Because the chip deforms during the cutting process, γ_c is not equal to 1.

$$\gamma_c = \frac{1}{\sqrt{1 + \frac{c_l}{f \cos \beta_e}}}$$

where c_l is the chip tool contact length, which depends on the chip and tool material.

The stress generated by the cutting force is given as

$$\sigma_{cert} = \frac{F_c \cos \alpha - F_t \sin \alpha}{A_s}$$

where A_s is the area over which the chip shears. Hence,

$$A_s = \frac{F_d}{\sin \alpha}$$

which is the cross-sectional area of the chip as projected in the shear plane.
In above equation, writing F_t in terms of F_c:

$$F_t = \frac{F_c \tan(\phi - \beta_e)}{\sin \alpha}$$

where ϕ is the friction angle,

$$\phi = \frac{\pi}{4} \alpha + \beta_e$$

Here the stress constraint is

$$\sigma_{cut} \geq \sigma_{str}$$

$$\sin \alpha \, F_c \cos \alpha - \tan (\phi - \beta_e) \geq \sigma_{str}$$

The cutting force F_c in terms of the hardness (BHN) of the tool is given as

$$F_c = 40(1000 \, F_d)^{0.802} \times \text{BHN} \times 0.456 \, (\varepsilon/50 \, 180/\pi)^{0.64}$$

where the lip angle, ε, is given in terms of effective rack angle (β_e) and clearance angle, ϕ,

$$\varepsilon = \frac{\pi}{2} - \beta_e - \theta$$

From Donaldson et al. (1973), the expression that approximates the maximum force F_{max} that a tool of nose radius r_n can withstand is

$$F_{max} = c_0 - c_1 - c_2 \cdot r_n^2$$

This nose radius constraint is $F_c \leq F_{max}$. Also, the heat dissipated from the chip is given as the product of the increase in chip temperature, ΔT.

The amount of heat produced per minute of metal cutting is:

$$H_e = \frac{F \, V_c}{\omega}$$

where ω is the mechanical equivalent of heat. Now, the heat constraint and the feed constraint are, respectively,

$$\eta \, H_e \leq H_s$$

$$r_n \geq 3F$$

4.5.4 Design Optimization Problem of the Single-Point Cutting Tool

Minimize $F_d\ (\beta_s,\ r_n) = \gamma$ (shear stress)
Subject to

$g1$: $F_c \leq F_{max}$ (cutting force constraint)
$g2$: $\sigma_{str}\cdot F_d \leq \sin\alpha\cdot F_c\ (\cos\ (\alpha) - \tan\ (\alpha - \beta_e))$ (stress constraint)
$g3$: $\eta\ H_e \leq H_s \leq 0$ (heat constraint)
$g4$: $3F \leq r_n$ (nose radius constraint)

4.5.5 Implementation of GA

Variable Boundaries:

$$0.215 \leq \beta_s \leq 0.3$$

$$0.05 \leq r_n \leq 0.625$$

4.5.6 Constant Values

$\delta = 0.209$, rad.
$\theta = 0.107$ rad.
$\sigma_{str} = 80{,}000.0$ lb /sq. in
$c = 0.116$ BTU/ lb. °C.
$\eta = 70\ \%$
$\ni = 778.0$ ft. lb /BTU.
$\rho = 0.314$ lb/ in^3.
$P_{max} = 1000.K$ lb. ft / min.
$c_1 = 0.02$ in.
$T = 415$ °C.
$W_e = 0.03$ in^2.
BHN $= 215.0$
$n_1, n_2, n_3, C_1 = 0.05, 0.21, 0.02, 100.45.$
$d_i(i = 0,...2) = 0.704, -252.0, 61384.0.$
$e_i(i = 0,...2) = 140.54, -44.107.$
$g_i(i = 0,...2) = 0.0074, 2.629.$

4.5.6.1 Genetic Operators

Sample size:100
Number of generations:100
Binary coding:10 bits for each variables (total of 20 bits)
Selection operator: Tournament

Crossover probability (P_c): 0.7
Mutation probability (P_m): 0.02
The solution obtained from GA is 1.49.

4.5.7 COMPARISON OF RESULTS WITH SOLUTION OBTAINED BY GAME THEORY

Variables	Game theory results
β_s	0.29
r_n	0.625
β_b	0.15
F	0.03
d	0.15
v	224.0
F_d	1.51

Solution:

Variables	GTR	GA
β_s	0.29	0.295
r_n	0.625	0.625
F_d	1.51	1.49

The results obtained by GA are comparable with game theory results.

REFERENCES

Arora, J., *Introduction to Optimum Design,* McGraw-Hill International, Singapore, 1989.

Bras, B.A. and Mistree, F., Designing design processes in decision-based concurrent engineering, *SAE Transactions, Journal of Materials and Manufacturing,* 100, 5, 451–458, 1991.

Deb, K., *Multi-Objective Optimization Using Evolutionary Algorithms,* John Wiley & Sons, West Sussex, 2001.

Deb, K., *Optimization for Engineeering Design: Algorithms and Examples,* Prentice Hall of India (Pvt.) Ltd., New Delhi 1995.

Deb, K. and Goyal, M., A combined genetic adaptive search (GeneAS) for engineering design, *Computer Science and Informatics,* 26 (4), pp.30–45, 1996.

Donaldson, C., Le Cain, C.H. and Goold, V.C., *Tool Design,* McGraw-Hill, New York, 1973.

Dudley, D.W., *Practical Gear Design,* McGraw-Hill, New York, 1954.

Dudley, D.W., *Practical Gear Design: The Design, Manufacture, and Application of Gears,* McGraw-Hill, New York, 1962

Faculty of Mechanical Engineering Department, "Design data," PSG College of Technology, Coimbatore, India, 1985.

Iwata, K. et al., Optimization of cutting conditions for multi-pass operations considering probabilistic nature in machining processes, *Journal of Engineering for Industry*, 99, 210–217, 1977.

Johnson, R.C., *Optimal Design of Machine Elements,* John Wiley & Sons, New York, 1961.

Karandikar, H.M. and Mistree, F., The integration of information from design and manufacture through decision support problems, *Applied Mechanics Reviews,* 44(10), 150–159, 1991.

Pakala, R. and Jagannatha Rao, J.R., Study of concurrent decision-making protocols in the design of a metal cutting tool using monotonicity arguments, Systems Design Laboratory, Department of Mechanical Engineering, University of Houston, Houston, TX, http://www.egr.uh.edu/me/research/sdl/research/rama-tool-94.ps.

Papalambros, P.Y. and Wilde, D.G., *Principles of Optimal Design,* Cambridge University Press, New York, 1988.

Shaw, M.C., *Metal Cutting Principles,* Clarendon Press, Oxford, 1984.

Taraman, S.R., and Taraman, K.S., Optimum selection of machining and cutting tool variables, *Proceedings of the First International Material Removal Conference,* Detroit, MI, SME Technical Paper MR83-182, 1983.

5 Optimization of Machining Tolerance Allocation

5.1 DIMENSIONS AND TOLERANCES

Dimensioning and tolerance specifies engineering design requirements for function and relationships of part features. Size and location of features are determined by a dimension. Tolerance is the amount of variation permitted in the dimension; all dimensions have tolerance. A dimension is a joint number, which includes a basic dimension value with its tolerance. A tolerance considers functional engineering and manufacturing requirements.

5.1.1 CLASSIFICATION OF TOLERANCE

5.1.1.1 Tolerance Schemes

Parametric and geometric tolerances are the two types of tolerancing schemes. Parametric tolerance consists of identifying a set of parameters and assigning limits to the parameters that define a range of values. A typical example of parametric tolerancing is the "plus/minus" tolerancing. Geometric tolerancing assigns values to certain attributes of a feature, such as forms, orientations, locations, run-outs and profiles.

5.1.1.2 Tolerance Modeling and Representation

Although geometric tolerance addresses the weakness and intrinsic ambiguities of parametric tolerancing, it still poses its own weakness, due mainly to its informal way of defining the core concepts. A branch of research that seeks an efficient way of defining and representing the tolerance information mathematically or electronically is called tolerance modeling and representation.

5.1.1.3 Tolerance Specification

Tolerance specification is concerned with how to specify tolerance "types" and "values." In practice, the tolerances are specified by the designer, based mainly on experience or empirical information. Tolerance specification is carried out preferably in conformance with the tolerance standards (International, ISO 1101, ANSI Y4.5, or company specific).

5.1.1.4   Tolerance Analysis

Tolerance analysis is a production function and is performed after the parts are in production. It involves first, gathering data on the individual component variations; second, creating an assembly model to identify which dimensions contribute to the final assembly dimensions; and third, applying the measured component variations to the model to predict the assembly dimension variations. A defective assembly is one for which the component variations accumulate and exceed the specified assembly tolerance limits.

One method to verify the proper functionality of a design is taking into account the variability of the individual parts. While the methods of analysis can be either deterministic or statistical, the design models to be analyzed can be one-, two-, or three-dimensional.

5.1.1.5   Tolerance Synthesis

Tolerance synthesis, also called tolerance allocation, is carried out in a direction opposite to tolerance analysis, from the tolerance of the function of interest to the individual tolerances. It tries to complete the tolerance specification, originally made from experience or empirical knowledge, by incorporating some heuristic, optimization, and other methods.

5.1.1.6   Tolerance Transfer

Tolerance transer aims to transfer tolerance requirements from the design stage to the manufacturing stage, while respecting the precision capabilities of machinery, making full use of the allowance by the design and taking into account machining errors, set-up errors, tool wears, and so on. One of the most classical and widely used techniques for the tolerance transfer is the tolerance charting method, but this is restricted basically to the one-dimensional case.

5.1.1.7   Tolerance Evaluation

Tolerance evaluation deals with how to assess the geometric deviations of a part using the data obtained from the coordinate measuring machines.

5.1.2   Tolerance and Cost Relationship

Earlier research on tolerance synthesis focused mainly on the formulations of a tolerance assignment as an unconstrained optimization problem and their close-form solutions. Based upon the general characteristics of a manufacturing cost-tolerance data curve, several general cost-tolerance relation models, including the exponential, reciprocal-squared and the reciprocal-powers models, were introduced. In addition, it fails to consider the valid range of a cost-tolerance curve to avoid infeasible solutions and requires manual formulation.

Cost-based optimal tolerance analysis techniques are very helpful in promoting economic design for functionality. They require a good deal of insight into developing a proper math model that relates cost and functional quality; once such a model is properly defined, the power of this optimal design tolerance becomes quite evident. This work has concentrated on minimizing manufacturing costs and minimizing the quality loss value.

Tolerance must be linked to more than the variability that originates in the manufacturing environment; it must have some costs that are incurred to make the product. These costs are primarily represented by the term unit manufacturing cost (UMC). Tolerances must be further developed in the context of two more costs. The life cycle cost (LCC) of the design will account for the broader quantity, which is repair and replacement cost associated with the use of the product. This metric is particularly important in industries that must repair and service the products they sell to satisfy customer expectations. Quality loss function (QLF) means that when the customer's requirement could not be reached by design parameters or deviated from the target point due to some manufacturing constraints, the manufacturer must pay for these expenses.

The tighter the tolerance, the more expensive it becomes to manufacture a part. This trend provides a fundamental rule in selecting tolerances by the designers at the design phase: tolerances should be chosen as large as possible as long as they meet the functional and assembly requirements of the part. Changing designs may be worthwhile to relax tolerance requirements for cost purposes. Larger tolerances result in using less accurate machines, lower inspection costs, and reduced rejection of material.

5.1.3 Quality Loss Function

Companies that practice on-target engineering use an alternative approach to the limitations of the step function exhibits as a measure of quality. The quality loss function was developed by Taguchi to provide a better estimate of the monetary loss incurred by manufacturers and consumers as product performance deviates from its target value. The quality loss function can be shown as Equation (5.1).

$$L(y) = k(y - m)^2 \qquad (5.1)$$

where $L(y)$ is the loss of money due to a deviation away from targeted performance as a function of measured response y of product; m is the target value of the product's response and k is an economic constant called the quality loss coefficient. The drawing is shown in Figure 5.1.

At $y = m$, the loss is zero and it increases the further y deviates from m. The quality loss curve typically represents the quality loss for an average group of customers. The quality loss for a specific customer would vary depending on the

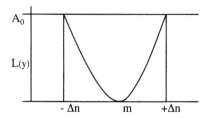

FIGURE 5.1 Illustrates the quality loss function.

the customer's tolerance and usage environment. However, deriving an exact loss function is not necessary for all situations. That practice would be too difficult and not generally applicable. The quality loss function can be viewed on several levels:

- As a unifying concept of quality and cost that allows one to practice the underlying philosophy driving target engineering.
- As a function that allows the relation of economic and engineering terms in one model.
- As an equation that allows detailed optimization of all cost, explicit and implicit, incurred by the firm, customers, and society through the production and use of a product.

5.1.4 TOLERANCE ALLOCATION METHODS

Tolerance allocation is a design function. It is performed early in the product development cycle, before any parts have been produced or tooling ordered. It involves first, deciding what tolerance limits to place on the critical clearances and fits for an assembly, based on performance requirements; second, creating an assembly model to identify which dimensions contribute to the final assembly dimensions; and third, deciding how much of the assembly tolerance to assign to each of the contributing components in the assembly.

A defective assembly is one for which the component variations accumulate and exceed the specified assembly tolerance limits. The yield of an assembly process is the percent of assemblies that are not defective. In tolerance analysis, component variations are analyzed to predict how many assemblies will be within specifications. If the yield is too low, rework, shimming or parts replacement may be required.

In tolerance allocation, an acceptable yield of the process is first specified and component tolerances are then selected to assure that the specified yield will be met. Often, tolerance design is performed by repeated application of tolerance analysis using trial values of the component tolerances. However, a number of algorithms have been proposed for assigning tolerances on a rational basis, without resorting to trial and error. The tolerance allocation methods can be classified based on the design rule and search rule, illustrated in Figure 5.2.

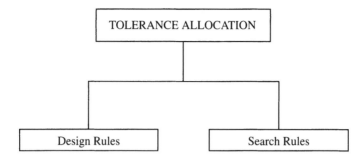

FIGURE 5.2 Tolerance allocation methods.

5.1.5 PROPORTIONAL SCALING METHOD

The designer begins by assigning reasonable component tolerances based on process design guidelines. He or she then sums the component tolerances by a constant proportionality factor. In this procedure, the relative magnitude of the component tolerances are preserved.

This method is demonstrated graphically in Figure 5.3 for an assembly tolerance T_{asm}, which is the sum of two component tolerances, T_1 and T_2. The straight line labeled as the "Worst Case Limit" is the locus of all possible combinations

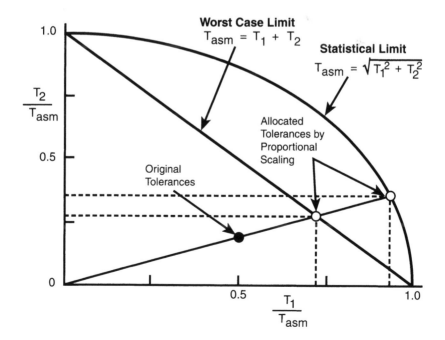

FIGURE 5.3 Graphical interpretation of tolerance allocation by proportional scaling.

of T_1 and T_2 that, added linearly, equal T_{asm}. The ellipse labeled "Statistical Limit" is the locus of root sum squares of T_1 and T_2 that equal T_{asm}. The following equations describe these two cases.

5.1.5.1 Worst Case Limit

$$T_{asm} = T_1 + T_2 + T_3 + \cdots T_n \qquad (5.2)$$

5.1.5.2 Statistical Limit

$$T_{asm} = \sqrt{T_1^2 + T_2^2 + T_3^2 + \cdots T_n^2} \qquad (5.3)$$

5.1.6 ALLOCATION BY WEIGHT FACTORS

A more versatile method of assigning tolerances is by weight factors, W. Using this algorithm, the designer assigns weight factors to each tolerance in the chain and the system distributes a corresponding fraction of the tolerance pool to each component. A larger weight factor W for a given component means a larger fraction of the tolerance pool will be allocated to it.

In this way, more tolerance can be given to those dimensions that are the more costly or difficult to hold, thus improving the producibility of the design. Figure 5.4 illustrates this algorithm graphically for a two-component assembly.

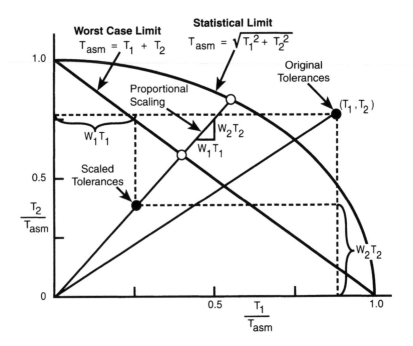

FIGURE 5.4 Graphical interpretation of tolerance allocation by weight factors.

The original values for component tolerances T_1 and T_2 are selected from process considerations and are represented as a point in the figure, as before. The tolerances are scaled, similar to proportional scaling; only the scale factor is weighted for each component tolerance, so the greater scale factors yield, the least reduction in tolerance.

5.1.6.1 Worst Case Limit

$$T_{asm} = W_1 T_1 + W_2 T_2 + W_3 T_3 + \cdots W_n T_n \tag{5.4}$$

5.1.6.2 Statistical Limit

$$T_{asm} = \sqrt{W_1 T_1^2 + W_2 T_2^2 + W_3 T_3^2 + \cdots W_n T_n^2} \tag{5.5}$$

5.1.7 Constant Precision Factor Method

Parts machined to a similar precision will have equal tolerances only if they are of equal sizes. As part size increases, tolerance (T_i) generally increases approximately with the cube root of size.

$$T_i = P(D_i)^{1/3} \tag{5.6}$$

where
  D_i = Basic size of the part
  P = Precision factor

Based on this rule, the tolerance can be distributed accordingly to the part size. The precision factor method is similar to the proportional scaling method except no initial allocation is required by the designer. Instead, the tolerances are initially allocated according to the nominal size of each component dimension and then scaled to meet the specified assembly tolerance.

5.1.8 Taguchi Method

The Taguchi method not only determines tolerance, but also determines the ideal nominal values for the dimensions. The method finds the nominal dimensions that allow the largest, lowest-cost tolerances to be assigned. It selects dimensions and tolerance with regard to their effect on a single design function. The method uses fractional factorial experiments to find the nominal dimensions and tolerance that maximize the so-called "signal-to-noise" ratio. The "signal" is a measure of how close the design function is to its desired nominal value. The "noise" is a measure of the variability of the design function caused by tolerances.

The main disadvantage of the Taguchi method is its inability to handle more than one design function. Finding one design function for a product might not be practical.

5.1.9 Tolerance Allocation Using Least Cost Optimization

A promising method of tolerance allocation uses optimization techniques to assign component tolerances that minimize the cost of production of an assembly. This is accomplished by defining a cost-tolerance curve for each component part in the assembly. The optimization algorithm varies the tolerance for each component and searches systematically for the combination of tolerances that minimize the cost.

Figure 5.5 illustrates the concept simply for a three-component assembly. Three cost-tolerance curves are shown. Three tolerances (T_1, T_2 and T_3) are initially selected. The corresponding cost of production is $C_1 + C_2 + C_3$. The optimization algorithm tries to increase the tolerances to reduce cost; however, the specified assembly tolerance limits the tolerance size. If tolerance T_1 is increased, then tolerance T_2 or T_3 must decrease to keep from violating the assembly tolerance constraint. Determining by inspection which combination will be optimum is difficult, but from the figure a decrease in T_2 results in a significant decrease in cost while a correponding decrease in T_3 results in a smaller increase

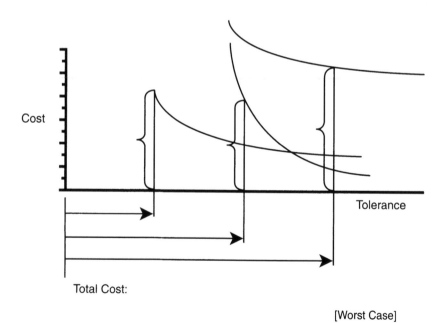

FIGURE 5.5 Optimal tolerance allocations for minimum cost.

in cost. In this manner, tolerances could be manually adjusted until no further cost reduction is achieved. The optimization algorithm is designed to find this optimum with a minimum of iteration. Note that the values of the set of optimum tolerances will be different when the tolerances are summed statistically than when they are summed by worst case.

5.1.10 TOLERANCE ANALYSIS VERSUS TOLERANCE ALLOCATION

The analytical modeling of assemblies provides a quantitative basis for the evaluation of design variations and specification of tolerances. An important distinction in tolerance specification is that engineers are more commonly faced with the problem of tolerance allocation rather than tolerance analysis.

The difference between these two problems is illustrated in Figure 5.6. In tolerance analysis, the component tolerances are all known or specified and the resulting assembly variation is calculated. On the other hand, in tolerance allocation, the assembly tolerance is known from design requirements, although the magnitudes of the component tolerances to meet these requirements are unknown. The available assembly tolerance must be distributed or allocated among the components in some rational way. The influence of the tolerance accumulation model and the allocation rule chosen by the designer on the resulting tolerance allocation will be demonstrated.

Another difference in the two problems is the yield or acceptance fraction of the assembly process. The assembly yield is the quality level. It is the percent of assemblies that meet the engineering tolerance requirements and may be expressed as the percent of acceptable assemblies or percent of rejects. For high quality levels, the rejects can be expressed in parts-per-million (ppm), that is, the number of rejects per million assemblies.

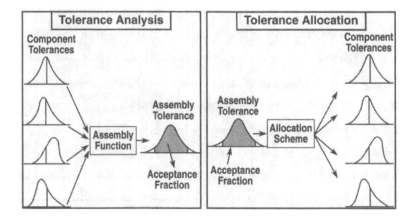

FIGURE 5.6 Tolerance analysis versus tolerance allocation.

5.1.11 TOLERANCE DESIGN OPTIMIZATION

The component tolerance could be distributed equally among all of the parts in an assembly. However, each component tolerance may have a different manufacturing cost associated due to part complexity or process differences. By defining a cost-tolerance function for each component dimension, the component tolerances may be allocated to minimize cost of production. The flow chart of a general design optimization process is given in Figure 5.7.

In some situations, two or more cost functions may exist. This situation is called a multiple objective optimization problem. The feasible region is the set of all solutions to the problem satisfying all the constraints. The optimal solution for a minimization problem is the solution with the smallest cost value in the feasible region.

In tolerance design optimization, the decision parameters are the tolerances of the components of an assembly. The cost function is a combined objective function of manufacturing cost and penalty cost. In this work, an unconventional

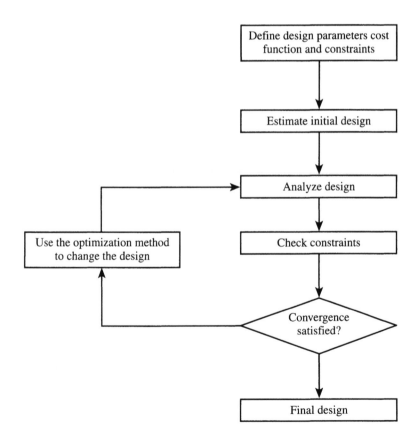

FIGURE 5.7 Tolerance design optimization process.

optimization method is used to find the optimal tolerance value. The final optimal tolerance value is achieved when the manufacturing cost is a minimum and the quality loss cost becomes zero.

5.1.12 NEED FOR OPTIMIZATION

Optimization algorithms are becoming increasingly popular in engineering design activities, primarily because of the availability and affordability of high-speed computers. They are extensively used in those engineering problems where the emphasis is on maximizing or minimizing a certain goal.

For example, optimization is routinely used in aerospace deign activities to minimize the overall weight of the aircraft. Thus, the minimization of the weight of the aircraft components is of major concern to aerospace designers. Chemical engineers, on the other hand, are interested in designing and operating a process plant for an optimum rate of production. Mechanical engineers design mechanical components for the purpose of achieving either a minimum manufacturing cost or a maximum component life. Production engineers are interested in designing an optimum schedule of the various machining operations to minimize the ideal time of machines and the overall job completion time. Civil engineers are involved in designing buildings, bridges, dams, and other structures to achieve a minimum overall cost, or maximize safety, or both. Electrical engineers are interested in designing communication networks to achieve minimum time for communication from one node to another.

All the above-mentioned tasks using either minimization or maximization (collectively known as optimization) of an objective require knowledge about the working principles of different optimization methods.

5.2 TOLERANCE ALLOCATION OF WELDED ASSEMBLY

5.2.1 PROBLEM STATEMENT

Assemblies and subassemblies are formed by combining two or more compo-nents. The dimension of interest in an assembly may be the sum or the difference of the individual components. The problem considers an assembly made of four components (Figure 5.8) welded together, with the weld thickness being neg-ligible. The quality characteristics of interest are the length of the assembly, Y, denoting the length of the components A, B, C, and D as X_1, X_2, X_3, and X_4, respectively. The length of the assembly is expressed as

$$Y = X_1 + X_2 + X_3 + X_4$$

In general, the dimension of interest is expressed as a linear combination of some individual component dimensions. Based on the process variance, the variation

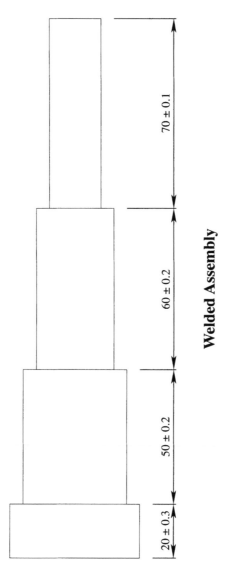

FIGURE 5.8 Welded assembly.

of dimension Y can be expressed in terms of the standard deviation of individual components:

$$\sigma^2 Y = \sigma^2_{X1} + \sigma^2_{X2} + \sigma^2_{X3} + \sigma^2_{X4}$$

From the above relationship, σ_Y can be obtained. So, the assembly tolerance of dimension Y is $\pm\, 3\sigma_Y$.

The mean length of the four components and their respective tolerance are given below:

Components	Mean Length (mm)	Tolerance (mm)
A	20	± 3
B	50	± 2
C	60	± 2
D	70	± 1

The above problem can be formulated as an optimization problem to get zero percentage rejection, i.e., allocating the individual tolerances for getting the required assembly tolerance. For example, take the required assembly tolerance as 200 ± 3 mm.

5.2.2 IMPLEMENTATION OF GA

5.2.2.1 Coding Scheme

The variables are coded with 20-digit binary numbers (each variable equals 5 digits).

Code	01010	00100	10001	10110
Decode	10	04	17	22
Variable (Decode × 0.1)	1	0.4	1.7	2.2

5.2.2.2 Objective Function

Zero percentage rejection, i.e., if the assembly tolerance is equal to the required tolerance of ± 3 mm, then there is no rejection.

5.2.2.3 Evaluation

To meet zero percentage rejection, the error value is calculated as follows:

$$\text{Total error} = (\text{Required tolerance} - \text{obtained tolerance})^2$$

5.2.2.4 GA Parameters

Sample size: 30
Number of generations: 100
Reproduction operator: Rank selection method
Crossover probability (P_c): 0.6
Mutation probability (P_m): 0.01

5.2.3 RANK SELECTION METHOD

The rank selection method is used for reproduction. The individuals in the population are ranked according to fitness and the expected value of each individual depends on its rank rather than on its absolute fitness. Ranking avoids giving the largest share of offspring to a small group of highly fit individuals, and thus reduces the selection pressure when the fitness variance is high. It also keeps selection pressure up when the fitness variance is low: the ratio of expected values of individuals ranked i and $i + 1$ will be the same whether their absolute fitness differences are high or low.

The linear ranking method proposed by Baker is as follows: each individual in the population is ranked in increasing order of fitness, from 1 to N. The expected value of each individual i in the population at time t is given by

$$\text{Expected value } (i,t) = \text{Min} + (\text{Max} - \text{Min}) \frac{rank(i,t) - 1}{N - 1}$$

where $N = 30$, Max $= 1.6$, and Min $= 0.4$. After calculating the expected value of each rank, reproduction is performed using Monte Carlo simulation by employing random numbers.

Probability of selection obtained from the expected value (Expected value/N) and the simulation run is given in Table 5.1.

5.2.4 OPTIMIZATION RESULTS

	A	B	C	D	Assembly
Obtained tolerance	1.1	2	2	0.1	3

5.3 TOLERANCE DESIGN OPTIMIZATION OF OVERRUNNING CLUTCH ASSEMBLY

5.3.1 PROBLEM DEFINITION

5.3.1.1 Optimum Tolerances for Overrunning Clutch

The overrunning clutch model was proposed by Feng and Kusiak in 1997. The overrunning clutch given in Figure 5.9 consists of three components: hub, roller, and cage. The contact angle Y is the functional dimension that must be controlled with the tolerance stackup limit and is expressed as an equation. The cost tolerance data for the clutch (tolerance in 10^4 inches, cost in dollars) is given in Table 5.2.

$$Y = f(X_1, X_2, X_3) = a \cos\left(\frac{X_1 + X_2}{X_3 - X_2}\right)$$

where a is constant. The nominal values of the three components of the overrunning clutch are

TABLE 5.1
Rank Selection Data

Rank	Probability of Selection	Cumulative Probability	Random Number	Selected Rank
1	0.0133	0.0133	0.218	11
2	0.0147	0.023	0.112	7
3	0.0160	0.044	0.711	25
4	0.0174	0.0614	0.655	24
5	0.0188	0.0802	0.419	18
6	0.0202	0.1004	0.354	16
7	0.0216	0.122	0.174	10
8	0.0229	0.1449	0.910	29
9	0.0243	0.1692	0.076	5
10	0.0257	0.1949	0.249	12
11	0.0271	0.222	0.129	8
12	0.0285	0.2505	0.439	18
13	0.0298	0.2803	0.380	17
14	0.0312	0.3115	0.498	20
15	0.0326	0.3441	0.134	8
16	0.0340	0.3781	0.159	9
17	0.0354	0.4135	0.966	30
18	0.0367	0.4502	0.761	26
19	0.0381	0.4883	0.850	28
20	0.0395	0.5278	0.697	24
21	0.0409	0.5687	0.579	22
22	0.0422	0.6109	0.636	23
23	0.0436	0.6545	0.416	18
24	0.0450	0.6995	0.035	3
25	0.0464	0.7459	0.913	29
26	0.0478	0.7937	0.582	22
27	0.0491	0.8428	0.628	23
28	0.0505	0.8933	0.752	26
29	0.0519	0.9452	0.897	29
30	0.0533	1.0000	0.232	12

Hub X_1: 2.17706 in
Roll X_2: 0.90000 in
Cage X_3: 4.00000 in
Tolerance of angle Y: 0.122 ± 0.035 rad

Tolerance boundary values of these components of the overrunning clutch are

Hub X_1: 1 to 120 (10^{-4} in)
Roll X_2: 1 to 5 (10^{-4} in)
Cage X_3: 1 to 120 (10^{-4} in)

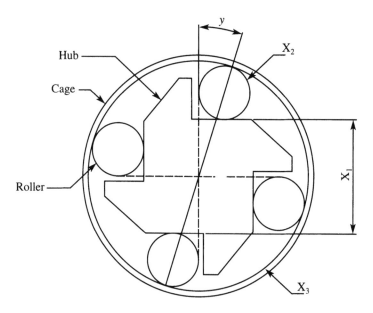

FIGURE 5.9 Overrunning clutch assembly.

5.3.1.2 Objective Function

Combined objective function is considered. It is the combination of minimizing the manufacturing cost and the cost associated with the quality loss function. The following manufacturing cost functions (m) are found in the literature for the manufacturing of hub, roller, and cage.

TABLE 5.2
Cost-Tolerance Data for the Clutch

Hub		Roll		Cage	
Tolerance	Cost	Tolerance	Cost	Tolerance	Cost
–	–	1	3.513	1	18.637
2	19.38	2	2.48	2	12.025
4	13.22	4	1.24	4	5.732
8	5.99	8	1.24	8	2.686
16	4.505	16	1.20	16	1.984
30	2.065	30	0.413	30	1.447
60	1.24	60	0.413	60	1.200
120	0.825	120	0.372	120	1.033

Tolerance in 10^{-4} inches; cost in dollars.

Manufacturing cost for single side tolerance values for

$$\text{Hub } M(t_1) = -0.731 + \frac{0.0580}{t_1^{0.688}}$$

$$\text{Roll } M(t_2) = -8.3884 + \frac{5.7807}{t_2^{0.0784}}$$

$$\text{Cage } M(t_3) = -0.978 + \frac{0.0018}{t_3}$$

The total manufacturing cost is

$$M(t_3) = M(t_1) + M(t_2) + M(t_3)$$

The cost associated with quality loss function is

$$Q(t_i) = \sum_{k=1}^{K} \frac{A}{T_k^2} \sigma_k^2$$

where
  A = Quality loss coefficient
  T_k = Single side functional tolerance stackup limit for dimensional chain k
  σ_k = Standard deviation of dimensional chain k
  K = Total number of the dimensional chain
  k = Dimensional chain index

From the above equations, the combined objective function can be formulated as

$$\text{Minimize } Y(t_1) = \sum_{i=0}^{3} [M(t_i) + Q(t_i)]$$

$$\text{Min } Y(t_1, t_2, t_3) = 33.3066 + \frac{0.058}{t_1^{0.688}} + 4 \times \frac{5.7807}{t_2^{0.0784}} + \frac{0.0018}{t_3}$$

$$+ \left[\frac{A}{(3)^2 \times (0.035)^2} \right] (t_1^2 + 4t_2^2 + t_3^2)$$

subject to

$$3.7499t_1 + 2 \times 7.472t_2 + 3.722t_3 \leq 0.0350$$

$$0.0001 \leq t_1 \leq 0.0120$$

$$0.0001 \leq t_2 \leq 0.0005$$

$$0.0001 \leq t_3 \leq 0.0120$$

5.3.2 IMPLEMENTATION OF PARTICLE SWARM OPTIMIZATION (PSO)

5.3.2.1 Coding System

To solve the problem using the random function, the tolerance values for each component of the overrunning clutch can be initialized. This can be called the initial population. Before that, the number of particles and number of iterations must be defined. These initial values are used to calculate the optimal cost by using the combined objective function. The overrunning clutch has three components, hub, roll, and cage; hence, the three variables in the objective function are t_1, t_2 and t_3, respectively. These three variables are initialized and the objective function is then calculated. This can be done for all particles used in the program. After that, find the present best (*pbest*) and global best (*gbest*) particles. By using these values, present particle velocities are updated for the next iteration. In the final iteration, the optimal cost and its optimal tolerance have been obtained.

5.3.2.2 Parameters Used

Number of particles: 10 to 30
Number of iterations: 50 to 500
Dimension of particles: 3
Range of each particle
Tolerance of hub: ≤ 0.0120 inches
Tolerance of roll: ≤ 0.00050 inches
Tolerance of cage: ≤ 0.0120 inches
Velocity of each particle
Velocity of hub: ≤ 0.0120 inches
Velocity of roll: ≤ 0.00050 inches
Velocity of cage: ≤ 0.0120 inches
Learning factors
C_1: 2
C_2: 2
Inertia weight factor (ω): 0.9

5.3.3 RESULTS AND DISCUSSION

The particle swarm optimization algorithm was run with a swarm size of 30, inertia weight factor of $\omega = 0.9$, and learning factors $C_1 = 2$ and $C_2 = 2$. The results

TABLE 5.3
Comparison of PSO Results with GA and GP

A	t₁	t₂	t₃	Y_{GP}	Y_{GA}	Y_{PSO}	Difference between Objective Function Value of PSO & GA	Difference between Objective Function Value of PSO & GP
	Tolerance by PSO (10^4 in)			Objective Function Value (U.S. Dollars)				
0	0.0049910	0.0005000	0.0023669	11.640	11.640	11.638	−0.002	−0.002
1	0.0050029	0.0005000	0.0024450	11.611	11.613	11.613	0	+0.002
52	0.0048399	0.0005000	0.0025192	11.793	11.789	11.785	−0.004	−0.008
100	0.0047619	0.0005000	0.0025939	11.918	11.923	11.920	−0.003	+0.002
300	0.0046619	0.0005000	0.0026985	12.466	12.467	12.466	−0.001	0
520	0.0042444	0.0005000	0.0026764	13.047	13.047	13.047	0	0

of PSO are compared with those obtained with genetic algorithm (GA) and geometric programming (GP) methods and are given in Table 5.3. Figure 5.10 shows the solution history of the result using PSO.

This comparison clearly concludes that the PSO technique yields the optimal machining tolerance allocation of the overrunning clutch assembly. The results are compared with GA and for the four values of A (0, 52, 100, and 300), an improvement in the combined objective function is achieved (0.002, 0.004, 0.003, and 0.001), and for the remaining two values of A (1 and 520), it matches exactly. Results are also compared with GP and for the two values of A (0 and 52), an improvement is achieved (0.002 and 0.008); for the two values of A (300 and 520), it matches exactly; and for the remaining two values of A (1 and 100), a marginal increase is obtained.

5.4 TOLERANCE DESIGN OPTIMIZATION OF STEPPED CONE PULLEY

The details of the stepped cone pulley is shown in Figure 5.11.

5.4.1 OBJECTIVE FUNCTION

- Minimization of total machining cost

5.4.2 DECISION VARIABLES

- Machining tolerances — continuous
- Machining datum — discrete

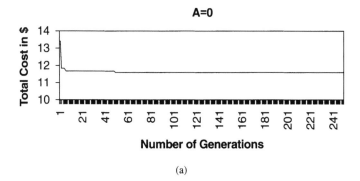

(a)

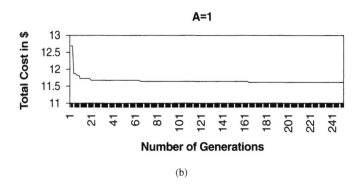

(b)

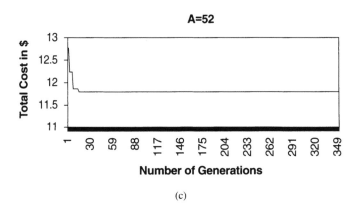

(c)

FIGURE 5.10 (a) Overrunning clutch assembly solution history using PSO (A = 0). (b) Solution history (A = 1). (c) Solution history (A = 52). (d) Solution history (A = 100). (e) Solution history (A = 300). (f) Solution history (A = 520).

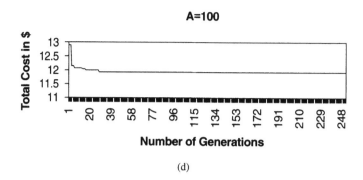

(d)

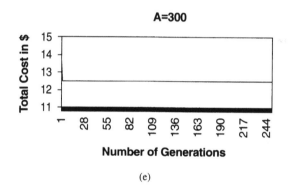

(e)

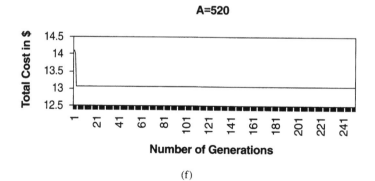

(f)

FIGURE 5.10 (continued).

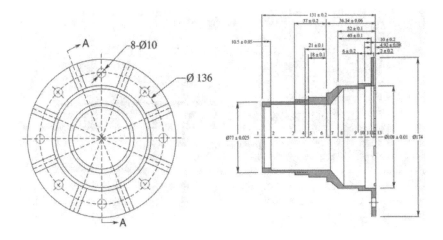

FIGURE 5.11 Stepped cone pulley.

5.4.3 Constraints

- Feasible machining datum
- Blue print tolerances
- Machining stock removal
- Machining tool capability

This problem can be formulated as a mixed variables (continuous and discrete), nonlinear, constrained optimization.

5.4.4 Finish Turning Datum Surface

- For machining surface 2 —11 or 13

5.4.5 Process Requirements

- Six machine tools — 3 lathes, 1 milling, 2 grinders
- Four external step surfaces machining — one internal step and one hole need to be machined

- Operation requirements and the datum constraints of the stepped bar are given in Table 5.4 and Table 5.5.

5.4.6 Evaluate Population

Mill 1Cost (C) = 3 + 0.017/T
Lathe 1C = 0.5 + 0.12/T
Lathe 2C = 1.5 + 0.14/T
Lathe 3C = 2.2 + 0.026/T
Grinder 1C = 3 + 0.017/T
Grinder 2C = 4 + 0.001/T

TABLE 5.4
Operation Requirements of Stepped Block

Operation	Machine	Step	Surf	SR	Operation	Machine	Step	Surf	SR
Rough turn	Lathe 1	28	1	4.0	Rough	Grinder 1	14	11	0.3
		27	1	4.0	Grind		13	6	0.3
		26	13	4.0			12	3	0.3
		25	7	4.0	Finish turn	Lathe 3	11	2	1.0
		24	11	4.0			10	7	1.0
		23	8	4.0	Milling	Mill 1	9	12	0.3
		22	6	4.0	Finish	Grinder 2	8	13	0.3
		21	3	1.0	Grind		7	10	0.3
		20	13	1.0			6	11	0.3
		19	7	1.3			5	9	0.3
		18	1	12			4	6	0.3
		17	8	1.0			3	5	0.3
		16	11	1.0			3	3	0.3
		15	6	1.0	Finish turn	Lathe 3	1	4	0.5

TABLE 5.5
Datum Constraint

Operation Number	Machining Surface	Feasible Datum	Operation Number	Machining Surface	Feasible Datum
RT	1	6, 11, 13	6 RG	11	6.9, 13
RT	13	1.5, 6, 9, 11, 12		6	11, 13
	7	13		3	1.6, 11, 13
RT	11	6, 9, 13	7	2	11, 13
	8	11, 13		7	13
	6	11, 13	8 M	12	11, 13
	3	3, 1.6, 11,1 3	9 FG	13	1.5, 6.9, 11, 12
SFT	13	1, 5, 6.9, 11, 12	10 FG	10	13
	7	13		11	6, 9, 13
SFT	1	6, 11, 13		9	6.9, 13
	8	11, 13		6	11, 13
	11	6, 9, 13		5	6, 13
	6	11,13		3	1.6, 11, 13
	3	1.6, 11, 13	11 FT	4	6

RT = rough turning; SFT = semi-finish turning; FT = finish turning, milling; RG = rough grinding;
FG = finish grinding.

5.4.7 PROPOSED GA PARAMETERS

Population size: 30
Number of generations: 500
Reproduction: Rank selection
Crossover probability (P_c): 0.6
Mutation probability (P_m): 0.02

5.4.8 MACHINING DATUM

If it is a rough grinding of surface 11, datum surfaces are 6, 9, or 13.

Code	Decode
0, 1, 2	6
3, 4, 5	9
6, 7, 8, 9	13

In this problem, 84 bits is required to represent all the machining tolerances (21 operations) and datum surfaces (maximum of 6).

5.4.9 INITIALIZE POPULATION

5.4.10 CODING SCHEME

Machining tolerance allocation
    e.g., finish turning.
    FT can produce a tolerance level of IT7 and IT9.
    As per ISO standard, the tolerance range is 21 to 52 mm.
    To represent the above, 5-digit binary number is required.

Code	Decode
00000	21
11111	52
11000	45

    Datum selection
    To represent the datum, integers can be used to reduce the length of the string (0 to 9).
    For finish turning surface 2, datum surfaces can be either 11 or 13.

5.5 TOLERANCE DESIGN OPTIMIZATION OF STEPPED BLOCK ASSEMBLY

The assembly of a stepped block given in Figure 5.12 is considered to illustrate the proposed approach.

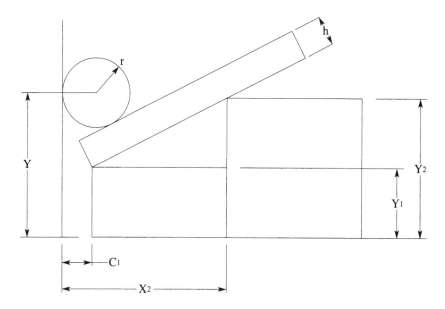

FIGURE 5.12 Stepped bar assembly.

5.5.1 Proposed Approach

In the following sections, two approaches for the conceptual design, sequential in nature, are presented.

Optimization of nominal values of noncritical dimensions.
Allocation of tolerances for the optimal nominal values based on cost-based tolerance models.

5.5.2 Optimization of Nominal Values of Noncritical Dimensions

Ullman and Zhang took a fresh look at the functional relationship that exists between the dependent assembly variable (Y) and the dependent component variables (X_i) through the familiar expression

$$Y = f(X_1, X_2, X_3 \cdots X_n)$$

The sensitivity is calculated by taking the partial derivative of the functional relationship between the critical (assembly) dimension and the noncritical (component) dimensions:

$$S_i = \partial_Y / \partial x_i$$

Therefore, each sensitivity coefficient can be calculated to express the unique relationship that each component has with respect to the effect of its change to the change in the assembly dimension. They have also defined another very common expression that represents the variance in the critical dimension as a sum of the component variances as attenuated by their respective sensitivity coefficients,

$$D_y^2 = (S_i \times dx_i)^2$$

where

S_i = Sensitivity of dimension x_i on Y
D_{Xi} = Deviation of dimension X_i
D_Y = Deviation of critical dimension

The best way to control critical dimension deviation is to make the noncritical dimensions insensitive by simply altering the value of $(S_i \times dx_i)$. This approach helps avoid the cost associated with quality improvement based on buying down variance by tolerance tightening.

5.5.3 Problem Formulation

The objective of this proposed approach is to minimize the deviation of the critical dimension by altering the nominal values of noncritical dimensions. Keeping the target value of the critical dimension at an exact value and keeping the tolerances of the noncritical dimensions the same can determine the set of nominal values for the noncritical dimensions that will give the minimum deviation of the critical dimension.

In this case, the target value of the critical dimension is considered as an equality constraint, and gives an allowable range to each noncritical dimension; bounds on these values serve as inequality constraints.

$$\text{Minimize } D_y = f(X_1, X_2, Y_1, Y_2, h, r)$$

$$\text{Subject to } Y = Y_{target}$$

$$X_1 < X_i < X_u$$

$$X_2 - X_i > 0$$

$$Y_2 - Y_1 > 0$$

where

X_L = Lower bound for variables X_1, X_2, Y_1, Y_2, h, r
X_u = Upper bound for variables X_1, X_2, Y_1, Y_2, h, r

In this case, the last two inequality constraints are necessary to keep the strain stepping upward, as shown in the assembly diagram.

5.5.3.1 Numerical Illustration

Step 1: Initially, the nominal values are assumed to be as shown in Table 5.1. The corresponding tolerance values can be obtained by experience or any other

TABLE 5.6
Tolerance Data of the Stepped
Block Assembly

Name	Value (mm)	Tolerance (mm)
X_1	100	0.07
X_2	260	0.23
Y_1	80	0.10
Y_2	180	0.17
h	36	0.06
r	50	0.04

method of allocation. Tolerance data is given in Table 5.6. Also, the critical dimension Y is assumed to be 166.67.

Step 2: The sensitivity for each noncritical dimension is calculated by partially differentiating the assembly equation with respect to the corresponding noncritical dimension.

$$Y = Y_1 + (r + h) \sec \theta + (r - X_1) \tan \theta$$

where

$$\tan \theta = \frac{Y_2 - Y_1}{X_2 - X_1}$$

$$\frac{\partial_y}{\partial_{x_1}} = (y_2 - y_1) \left[\frac{(r+h)\sec \theta. \tan \theta}{(x_2 - x_1)^2 + (y_2 - y_1)^2} + \frac{(r - x_2)}{(x_2 - x_1)^2} \right]$$

$$\frac{\partial_y}{\partial_{x_2}} = -(y_2 - y_1) \left[\frac{(r+h)\sec \theta. \tan \theta}{(x_2 - x_1)^2 + (y_2 - y_1)^2} + \frac{(r - x_1)}{(x_2 - x_1)^2} \right]$$

$$\frac{\partial_y}{\partial_{y_1}} = 1 \frac{1}{(x_2 - x_1)} \left[\frac{(r+h)\sec \theta. \tan \theta}{(x_2 - x_1)^2 + (y_2 - y_1)^2} + (r - x_1) \right]$$

$$\frac{\partial_y}{\partial_{y_2}} = 1 \frac{1}{(x_2 - x_1)} \left[\frac{(r+h)\sec \theta. \tan \theta}{(x_2 - x_1)^2 + (y_2 - y_1)^2} + (r - x_1) \right]$$

$$\frac{\partial_y}{\partial h} = \sec \theta$$

$$\frac{\partial_y}{\partial h} = \sec \theta + \tan \theta$$

Using the above relations, the sensitivities are determined for the noncritical dimensions.

TABLE 5.7
Sensitivity Analysis

Name	Value	Tolerance (dx_i)	Sensitivity (S_i)	Deviation (dY)
X_1	100	0.07	0.642266	0.044959
X_2	260	0.23	0.017266	0.003971
Y_1	80	0.10	1.027626	0.102763
Y_2	180	0.17	0.027626	0.004696
h	36	0.06	1.179248	0.070755
R	50	0.04	1.804248	0.072170

Step 3: The next step is to determine the deviation by multiplying the tolerance with the corresponding sensitivities. The sensitivity analysis is given in Table 5.7.

Step 4: Now the critical dimension deviation (D_Y) is calculated using Equation 5.6:

$$D_Y = 0.151109$$

Step 5: The optimization technique is now applied to get the set of values of noncritical dimensions, which gives the minimum critical dimension deviation. The continuous ants colony (CACO) algorithm is used here to find the minimum critical dimension deviation.

5.5.4 IMPLEMENTATION OF CONTINUOUS ANT COLONY OPTIMIZATION (CACO)

The flow chart of the proposed continuous ant colony optimization (CACO) is given in Figure 5.13. Initially, 20 solutions are randomly generated. The first 12 solutions are taken as superior solutions and the next 8 solutions are taken as inferior solutions, as shown in Table 5.8. The percentage distribution of ants for local and global search is given Figure 5.14.

5.5.5 RANDOMLY GENERATED SOLUTIONS IN ASCENDING ORDER

Based on the algorithm, the superior solution must undergo a local search and the inferior solutions are thoroughly checked by a global search.

5.5.6 GLOBAL SEARCH FOR INFERIOR SOLUTIONS

The global search is used to modify the solutions in the inferior region (Solution 13 to Solution 20). The steps involved are crossover, mutation, and trail diffusion.

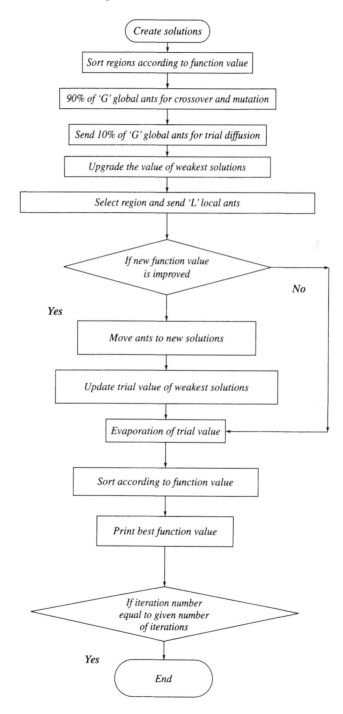

FIGURE 5.13 Flow chart of ant colony algorithm.

TABLE 5.8
Initial Random Solution

Solution	Number	Objective Function Value
Superior	1	0.131
	2	0.131
	3	0.138
	4	0.139
	5	0.149
	6	0.151
Inferior	7	0.151
	8	0.152
	9	0.156
	10	0.160
	11	0.160
	12	0.162
	13	0.162
	14	0.167
	15	0.168
	16	0.177
	17	0.186
	18	0.195
	19	0.222
	20	0.243

5.5.7 CROSSOVER OR RANDOM WALK

Crossover or random walk is a replacement strategy, usually done to modify the solutions by replacing the inferior solutions by superior region solutions. The following steps explain this procedure.

A random number is generated between 1 and 12. The corresponding solution in the superior region replaces the inferior solution.

Selected solutions in the superior region should be excluded so that they will not be selected again for replacement.

The crossover probability is taken as 0.9.

In case of Solution 13 (given in Table 5.9), the following steps are followed:

The random number 11 is generated.

The region 11 has the critical dimension deviation $D_Y = 0.159505$.

The Solution 13 given in Table 5.9 is replaced with the modified parameters given in Table 5.10.

The critical dimension deviation dy = 0.162333.

Similar procedure is carried out for the other inferior solutions from 14–18. Thus the random walk procedure is done.

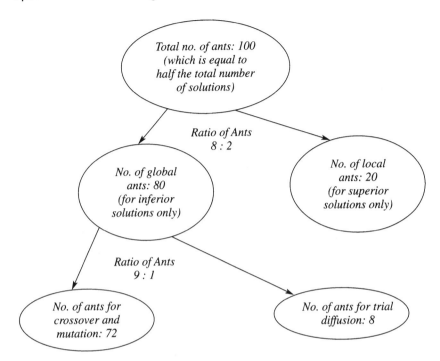

FIGURE 5.14 Distribution of ants for local and global search.

5.5.8 Mutation

Now, again the solutions from 13 to 18 must undergo mutation. In this step, the values of nominal dimensions are modified slightly according to

$$\Delta X = R[1 - r^{(1 - T)\,b}]$$

TABLE 5.9
Solution No. 13

Name	Value (mm)	Tolerance (mm)	Sensitivity	Deviation
X_1	99	0.07	0.772579	0.054081
X_2	247	0.23	0.078772	0.018118
Y_1	73	0.10	0.907474	0.090747
Y_2	199	0.17	0.092526	0.015729
h	34	0.06	1.313316	0.078799
r	55	0.04	2.164668	0.086587

TABLE 5.10
Modified Parameters of Solution 13

Name	Value (mm)	Tolerance (mm)	Sensitivity	Deviation
X_1	117	0.07	− 0.819260	0.057348
X_2	279	0.23	− 0.032592	0.007496
Y_1	84	0.10	0.961740	0.096174
Y_2	222	0.17	0.038260	0.006504
h	40	0.06	1.313641	0.078818
R	59	0.04	2.165493	0.086620

where the values are
b = a constant (obtained by trial)
T = the ratio of current iteration to the total number of iterations
r = random number
$R = (X_{max} - X_i)$

Here a mutation probability of 0.5 is set. Then, again, a random number is generated. If this random number is greater than 0.5, ΔX is then subtracted from the old value; otherwise, it is added to the old value. After replacement of the region 13 in the previous step, mutation is performed using the following steps.

The value of constant b is taken as 10.00.
The value of $T = 0.50$.
The random number (for example, 0.98) is generated.
The range of each parameter of this region is calculated using the expression,
$R = [X_{max} - X_i] = 40$.
Using r, R, T, and b; the value ΔX is calculated. The value of ΔX is found to be 3.843165.
To define whether ΔX must be positive or negative, the mutation probability is taken into consideration and here it is taken as 0.5.
Again, a random number is generated. If this random number is greater than 0.5, then Δx is subtracted from the old value; otherwise, it is added to the old value.
After modifying all the parameters, the critical dimension deviation has been found to be 0.182879.

The above-mentioned steps are performed on the remaining five solutions and the corresponding function values are determined.

5.5.9 TRAIL DIFFUSION

The remaining two Solutions 19 and 20 at the tail end are improved by the trail diffusion technique.

Two parent variables are selected from the superior region, for which two dissimilar random numbers are generated between 1 and 12 and named as Parent 1 and Parent 2.

A random number α is generated between 0 and 1 for every region. If α is less than or equal to 0.5, then

$$X_{child} = \alpha X_{parent} + (1 - \alpha)X_{Parent2}$$

If the random number is between 0.5 and 0.75, then

$$X_{child} = X_{Parent2}$$

For this new set of variables, the new solutions are complete. The two parents (Parent 1 and Parent 2) are randomly chosen from the superior region. In this case,

> Parent 1 (nominal value of X_1 from superior solution) = 80
> Parent 2 (nominal value of X_1 from superior solution) = 84

As in the case of Solution 19, the value of α ($\alpha = 0.543443$) lies between 0.5 and 0.75.

So Parent 1 is taken as new value and the other parameters are similarly modified. The objective function is then calculated to be 0.140774.

The same steps are followed to improve the other solutions.

5.5.10 LOCAL SEARCH

The local search improves the 12 solutions in the superior region only. The following steps explain the local search.

Initially the pheromone value (ph) for every region is set to 1.0 and the age for every region is taken as 10.

The average pheromone value ($ave\ ph$) is calculated using

$$ave\ ph = \frac{\sum ph}{\text{Number of superior solutions}}$$

For the first case, the average pheromone value will be 1.

A random number is then generated in the range 0 to 1.

If the random number is less than $ave\ ph$, the search is further pursued; else the ant quits, and then leaves the solution without any alteration.

Now the limiting step is calculated for the region using the constants k_1 and k_2. The k_1 and k_2 values are dependent upon the nature of problem.

$$\text{Limiting step: } (1s) = k_1 - (age \times k_2)$$

For this case, the k_1 and k_2 values are taken as $k_1 = 0.1$ and $k_2 = 0.001$. Here k_1 is always greater then k_2.

Again, a random number is generated and based on this random number, the following operation is performed:

$$X_{i,new} = X_i + \text{limiting step}$$

As the random number is greater than 0.5, the limiting step is added; else the limiting step is subtracted from the X_i value.

For the calculated variables, the new solution should now be calculated. Thus, the modified solution is obtained.

$$F(X_i) = F(X_{i,new})$$

The new average pheromone value is calculated as follows. If the current solution is less than the previous solution, the age is incremented by one; otherwise, it is decremented by one. The new pheromone value is calculated by using the following expression:

$$ph_{i,new} = \frac{F(X_{i,new}) - F(X_{i,old})}{F(X_{i,old})} + ph_{i,old}$$

The pheromone average (*ave ph*) now is calculated using the new values. The above steps are performed on the remaining 11 solutions and then 19 more such iterations are performed to improve the solutions in the superior region. In the case of Solution 1, the values are taken as $k_1 = 0.1$ and $k_2 = 0.001$. The limiting step is calculated to be 0.09. Also, now a random number is generated (for example, 0.6). The new value of X_1 is calculated to be 79.91. Similarly, the other parameters are also modified and the new solution is obtained. Since the new solution is inferior to the old solution, the average pheromone value of this solution is modified to $Ph_{i,new} = 0.998875$.

5.5.11 AFTER APPLYING CACO ALGORITHM

The values given in Table 5.11 are obtained after applying the optimization technique. The minimum critical dimension deviation is $D_Y = 0.125628$.

5.5.12 ALLOCATION OF TOLERANCES FOR OPTIMAL NOMINAL VALUES USING CACO

A promising method of tolerance allocation uses optimization techniques to assign component tolerances that minimize the cost of production of the assembly. This is accomplished by defining cost-tolerance curve for each component

TABLE 5.11
Optimization Results after Applying CACO Algorithm

Number	Name	Nominal Value (mm)	Tolerance (mm)	Sensitivity	Deviation
1	X_1	73	0.09	0.185498	0.012985
2	X_2	336	0.15	0.004496	0.001034
3	Y_1	70	0.09	0.976334	0.097633
4	Y_2	120	0.15	0.023666	0.004023
5	h	42	0.08	1.017889	0.061073
6	R	60	0.04	1.207884	0.048315

part in the assembly. The optimization algorithm varies the tolerance for each component and searches systematically for the combination of tolerances, which minimizes the cost. However, telling by inspection which combination will be optimum is difficult. The optimization algorithm is designed to find it with a minimum of iterations.

Figure 5.15 illustrates the concept simply for a three component assembly. Three cost-tolerance curves are shown. Three tolerances (T_1, T_2, and T_3) are initially selected. The corresponding cost of production is $C_1 + C_2 + C_3$. The optimization algorithm tries to increase the tolerances to reduce cost; however, the specified assembly tolerance limits the tolerance size. If tolerance T_1 is increased, tolerance T_2 or T_3 must then decrease to keep from violating the assembly tolerance constraint. Telling by inspection which combination will be optimum is difficult. But we can see from the figure that a decrease in T_2 results in a significant increase in cost, while a corresponding decrease in T_3 results in a smaller increase in cost. In this manner, tolerances could be manually adjusted until no further cost reduction is achieved. The optimization algorithm is designed to find the minimum cost automatically. Note that the values of the set of optimum tolerances will be different when the tolerances are summed statistically than when they are summed by worst-case limits.

5.5.13 PROBLEM FORMULATION

From the previous approach, the set nominal values can be obtained, which gives minimum critical dimension deviation. But with this approach, the cost-based optimal tolerances are allocated for the obtained nominal set values so that the cost of production of an assembly is minimized. The optimization algorithm randomly varies the tolerance for each component and searches the tolerances systematically for the combination of tolerances that minimizes the cost.

Therefore, the problem is multi-objective, i.e., minimization of total cost and minimization of critical dimension (assembly) function deviation. Since these criteria are on different scales, to reflect their real contribution to the multiple-criterion objective function, their values must be normalized to the same scale.

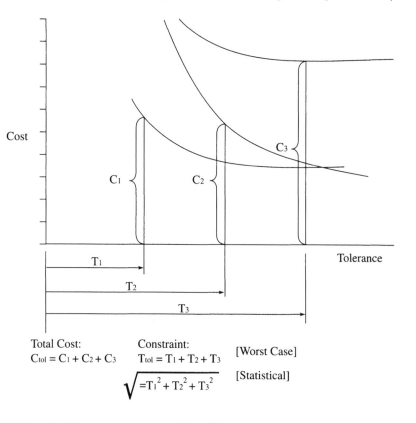

Total Cost: Constraint:
$C_{tol} = C_1 + C_2 + C_3$ $T_{tol} = T_1 + T_2 + T_3$ [Worst Case]

$$\sqrt{= T_1^2 + T_2^2 + T_3^2}$$ [Statistical]

FIGURE 5.15 Tolerance versus cost relationship.

5.5.14 MULTIPLE-CRITERION OBJECTIVE FUNCTION

The normalized objective value of a solution X, $V(X)$, can be as follows:

$$V(X) = \sum_{i=1}^{n} NW_i \times N(X_i)$$

where
  $V(X)$ = Normalized objective value of solution X
  W_i = Renormalized weight of criterion i
  NW_i = Normalized weight of criterion i

where

$$NW_i = \frac{W_i}{\sum_{i=2}^{n} W_i}$$

where

$$N(X_i) = \frac{mm_i}{X_i}$$

The above definitions of the multiple-criterion objective function are used to evaluate the combined objective functions.

5.5.15 OBJECTIVE FUNCTION

Minimization of assembly tolerance:

$$D_Y = \sqrt{\sum_{i=1}^{n} (S_i \times T_i)^2}$$

Subject to

$$Y = Y_{target}$$

$$X_L < X_i < X_u$$

$$X_2 - X_1 > 0 \text{ and } Y_2 - Y_1 > 0$$

Minimization of total cost:

$$C = \sum_{i=1}^{n} \left(A + \frac{B}{T_i^k} \right)$$

Subject to

$$t_1 \leq T_i \leq t_2$$

where
   A = Fixed cost (per part) in the process
   B = Cost of making a single component
   T = Tolerance and k depends on the process

The variables t_1 and t_2 are the minimum and maximum tolerances that can be produced for the concerned manufacturing process.

5.5.16 Results and Discussions

The quality of the product can be improved by different ways. Several publications and much research have been devoted to this type of situation. But the approach (i.e., reducing the sensitivity of the component dimensions by moving the nominal values to a less sensitive portion) is a useful method and is done in the concept design stage itself. Also, this approach does not require any additional cost to improve the quality.

The approach described in this work has proven a useful tool for improving the quality of the design during the concept development stage without any additional cost. The assignment of tolerances to the components of mechanical assemblies is then fundamentally an engineering problem. Therefore, in this work, a method is developed to automatically allocate the cost-based optimal tolerances to the components by considering a least cost model. In this approach, a continuous ant colony (CACO) algorithm is used as an optimization tool for both improving the quality of the design and allocating the optimal tolerances to the components. The method developed in this approach is used to improve the quality of the design with no additional costs. Also, this method will automatically allocate the optimal tolerances to the components in the mechanical assemblies.

The main limitations of this approach are that the variation of environmental and deteriorative effects on the assembly are not taken into consideration and that it may only apply to nonlinear mechanical assemblies. On the other hand, this approach will provide a much-needed practical method to improve the quality of the design and to allocate the tolerances of the components of nonlinear mechanical assemblies.

REFERENCES

Askin, R.G. and Goldberg, J.B., Economic optimization in product design, *Engineering Optimization,* 14, 139–152, 1988.

Carpinetti, L.C.R. and Chetwynd, D.G., Genetic search methods for assessing geometric tolerances, *Computer Methods in Applied Mechanics & Engineering,* 122, 193–204, 1995.

Chase, K.W., Greenwood, W.H., Loosli, B.G., and Hauglund, L.F, Least cost tolerance allocation form mechanical assemblies with automated process selection, *Manufacturing Review 3,* 49–59, 1990.

Choi Hoo-Gon, R., Park, M. H., and Selisbury, E. Optimal tolerance allocation with loss function, *Journal of Manufacturing Science and Engineering,* 122, 529–535, 2000.

Deb, K., *Optimization for Engineering Design: Algorithm and Examples,* Prentice Hall, New Delhi, India, 1995.

Greenwood, W. and Chase, K.W., Worst case tolerance analysis with nonlinear problems, *Journal of Engineering Industry,* 110, 232–235, 1988.

Lee, J. and Johnson, G.E., Optimal tolerance allotment using a genetic algorithm and truncated Monte Carlo simulation, *Computer Aided Design,* 25, 601–611, 1993.

Lee, W.J. and Woo, T.C., Optimization selection of discrete tolerance, *ASME Journal of Mechanism, Transmissions and Automation in Design,* III, 243–251, 1989.

Li, W., Bai, G., Zhang, C., and Wang, B., Optimization of machining datum selection and machining tolerance allocation with genetic algorithm, *International Journal of Production Research*, 38, 1407–1424, 2000.

Michael, W. and Siddall, J.N., Optimization problem with tolerance assignment and full acceptance, *ASME Journal of Mechanical Design*, 103, 842–848, 1981.

Partkinson, D.B., Assessment and optimization of dimensional tolerance, *Computer Aided Design*, 17(4), 191–199, 1985.

Speckhart, F.H., Calculation of tolerance based on a minimum cost approach, *ASME Journal of Engineering for Industry*, 94(2), 447–453, 1972.

Spotts, M.F., Allocation of tolerance to minimize cost of assembly, *ASME Journal of Engineering for Industry*, August, 762–764, 1973.

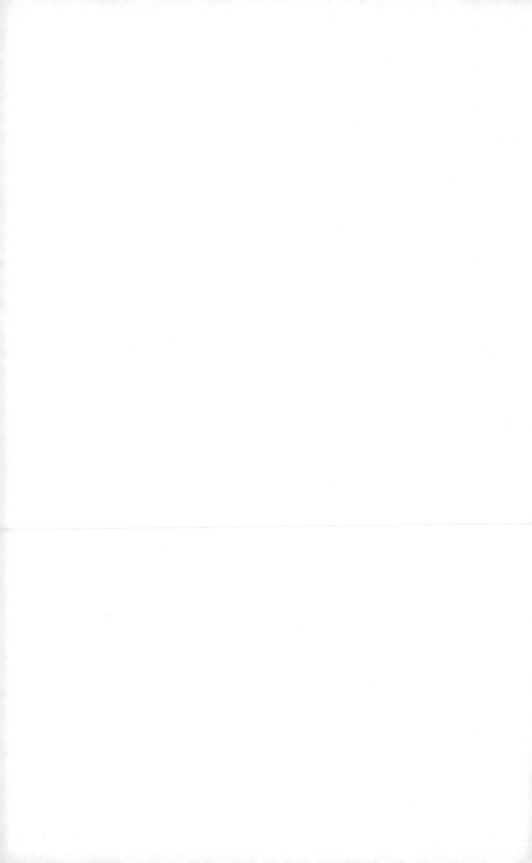

6 Optimization of Operating Parameters for CNC Machine Tools

Machining parameters optimization has significant practical importance, particularly for operating CNC machines. Due to the high cost of these machines, an economic need exists to operate them as efficiently as possible to obtain the required return on investment. Because the cost of machining on these machines is sensitive to the machining variables, the optimum values must be determined before a part is put into production. The operating parameters in this context are cutting speed, feed rate, depth of cut, and so on that do not violate any of the constraints that may apply to the process and satisfy the objective criterion, such as minimum time, minimum production cost, or maximum production rate.

6.1 OPTIMIZATION OF TURNING PROCESS

6.1.1 OBJECTIVE FUNCTION

Both the production cost and time are considered as objective functions. The production cost per component for a machining operation is comprised of the sum of the costs for tooling, machining, tool change time, handling time, and quick return time, which is given below:

$$C_u = C_o t_m + (t_m/T) \times (C_o t_{cs} + C_t) + C_o(t_h + t_R)$$

where the cutting time per pass is

$$t_m = D.L/1000.V.f$$

The total time required to machine a part is the sum of the times necessary for machining, tool changing, tool quick return, and workpiece handling.

$$T_u = t_m + t_{cs}(t_m/T) + t_R + t_h$$

Taylor's tool life equation is represented in terms of V, f, doc, and T:

$$V. f^{a1}. doc^{a2}. T^{a3} = K$$

where, a1, a2, a3 and K are the constants.

This equation is valid over a region of speed and feed by which the tool life (T) is obtained.

6.1.2 Nomenclature

D = Diameter of the workpiece (mm)
L = Length of the workpiece (mm)
V = Cutting speed (m/min)
f = Feed rate (mm/rev)
f_{min}, f_{max} = Minimum and maximum allowable feed rates
R_a = Surface roughness (μm)
$R_{a,max}(r)$, $R_{a,max}(f)$ = Maximum surface roughness of rough and finish cut, respectively
P = Power of the machine (kW)
F = Cutting force (N)
θ = Temperature of tool-workpiece interface (°C)
doc = Depth of cut (mm)
$doc_{min}(r)$, $doc_{max}(r)$ = Minimum and maximum allowable depth of cut (rough)
$doc_{min}(f)$, $doc_{max}(f)$ = Minimum and maximum allowable depth of cut (finish)
a_1, a_2, a_3, K = Constants used in tool life equation
T = Tool life (min)
t_m = Machining time (min)
t_{cs} = Tool change time (min/edge)
t_R = Quick return time (min/pass)
t_h = Loading and unloading time (min/pass)
T_u = Total production time (min)
C_o = Operating cost (Rs-rupees-currency/piece)
C_t = Tool cost per cutting edge (Rs/edge)
C_T = Total production cost (Rs/piece)

6.1.3 Operating Parameters

6.1.3.1 Feed Rate

The maximum allowable feed has a pronounced effect on both optimum spindle speed and production rate. Feed changes have a more significant impact on tool life than depth-of-cut changes. The system energy requirement reduces with feed because the optimum speed becomes lower. Therefore, the largest possible feed consistent with allowable machine power and surface finish is desirable for a machine to be fully utilized. Obtaining much higher metal removal rates without reducing tool life is often possible by increasing the feed and decreasing the speed. In general, the maximum feed in a roughing operation is limited by the force that the cutting tool, machine tool, workpiece, and fixture are able to withstand. The maximum feed in a finish operation is limited by the surface finish requirement and often can be predicted to a certain degree based on the surface finish and tool nose radius.

6.1.3.2 Cutting Speed

Cutting speed has a greater effect on tool life than either depth of cut or feed. When compared with depth of cut and feed, the cutting speed has only a secondary effect on chip breaking when it varies in the conventional speed ranges. Certain combinations of speed, feed, and depth of cut are preferred for easy chip removal and are dependent mainly on the type of tool and workpiece material. Charts providing the feasible region for chip breaking as a function of feed versus depth of cut are sometimes available by the tool manufacturers for a specific insert or tool and can be incorporated into the optimization systems.

6.1.4 Constraints

Maximum and minimum permissible feed rates, cutting speed, and depth of cut:

$$f_{min} \leq f \leq f_{max}$$

$$V_{min} \leq V \leq V_{max}$$

$$doc_{min} \leq doc \leq doc_{max}$$

Power limitation:

$$0.0373 \times V^{0.91} f^{0.78} doc^{0.75} \leq P_{max}$$

Surface roughness limitations, especially for a finish pass:

$$14{,}785 \times V^{1.52} f^{1.004} doc^{0.25} \leq R_{a,max}$$

Temperature constraint:

$$74.96 \times V^{0.4} f^{0.2} doc^{0.105} \leq \theta_{max}$$

Cutting force constraint:

$$844 \times V^{9.1013} f^{0.725} doc^{0.75} \leq F_{max}$$

The above constraints were taken from three independent sources and available in Agapiou (Part 1, 1992). They were empirically developed from experimental data collected specifically for other simulation studies (while turning plain carbon steel without coolant) and have an inherent uncertainty associated with their predictions. The values of the constants and the exponent coefficients are a function of the workpiece material, and the material and geometry of the tool. However, they were all obtained for the same boundaries of cutting speed, feed, and depth of cut. For the purpose of this problem, these equations are applicable. Also, their prediction changes during the life of the cutting tool and thus during the machining of each part because the machining process models are functionally related to the wear status of the tool. Hence, most of the constraint equations, and especially the surface finish model, should not be a function of only the speed, feed, and depth of cut but also should include the wear as an independent variable. However, experimental evaluation of tool life, even in particularly simple conditions, can require a great deal of testing due to the inherent scatter peculiar to tool wear processes. Nevertheless, similar constraints can be developed for a specific work material and tool geometry family under specific cutting parameters.

The determination of the constraint equations is a very tedious process but it does not require as extensive testing as for the tool life equation. However, the parameters of the constraints are very sensitive to a particular application. Changes in size or geometry of the part, machine, tool, coolant, and so on can all affect these parameters. Nevertheless, some of the constraints can be predicted within a 10 to 20% margin of error through analytical models developed based on the tool geometry, workpiece, and tool material properties and machine variables; while the size or geometry of the part and machine characteristics are incorporated as inputs. These inputs can be estimated reliably only by using generic cutting test data, such as that from turning tests. Therefore, the availability of either the turning test data or the empirical equations for the tool life and constraints should be advantageous for materials used in high volume year after year (i.e., gray cast iron or aluminum materials used for engine blocks and transmission bodies).

6.1.5 IMPLEMENTATION OF NELDER–MEAD SIMPLEX METHOD

6.1.5.1 Data of Problem

$L = 203$ mm
$D = 152$ mm
$V_{min} = 30$ m/min, $V_{max} = 200$ m/min
$f_{min} = 0.254$ mm/rev, $f_{max} = 0.762$ mm/rev
$R_{a,max}(r) = 12$ m, $R_{a,max}(f) = 8$ m
$P_{max} = 5$ kW
$F_{max} = 900$ N
$_{max} = 500°C$
$doc_{min}(r) = 2.0$ mm, $doc_{max}(r) = 5.0$ mm
$doc_{min}(f) = 0.6$ mm, $doc_{max}(f) = 1.5$ mm
$a_1 = 0.29$, $a_2 = 0.35$, $a_3 = 0.25$, $K = 193.3$
$t_{cw} = 0.5$ min/edge
$t_R = 0.13$ min/pass
$t_h = 1.5$ min/piece
$C_o =$ Rs. 3.50/min
$C_t =$ Rs. 17.50/edge

6.1.5.2 Solution by Nelder–Mead Simplex Method

The Nelder–Mead simplex method described in Chapter 2 is suitably modified for solving this problem. A flowchart of this method is shown in Figure 6.1.

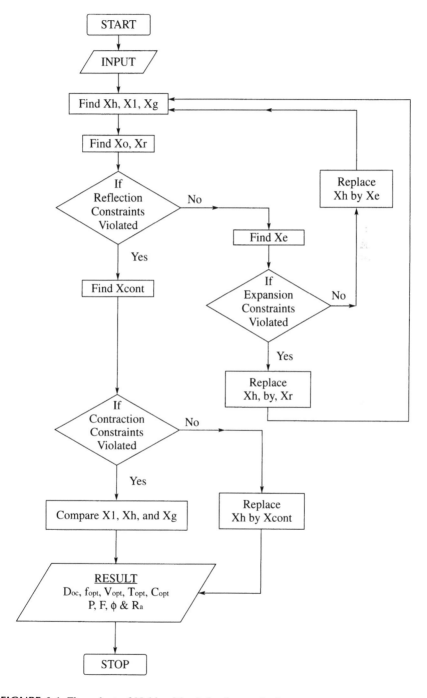

FIGURE 6.1 Flow chart of Nelder–Mead simplex method.

The initial simplex is formed by considering the minimum limit of speed and feed rate. The accuracy of the result is dependent upon the chosen initial simplex. The results are obtained for the following four simplexes and the best one is selected.

Simplex 1: (0.254, 30), (0.4, 30) and (0.4, 65)
Simplex 2: (0.254, 30), (0.4, 65) and (0.254, 45)
Simplex 3: (0.254, 30), (0.4, 30) and (0.254, 65)
Simplex 4: (0.254, 30), (0.4, 45) and (0.254, 65)
The results are given in the Table 6.1 and Table 6.2.

TABLE 6.1
Optimization of Single-Pass Turning Using Nelder–Mead Simplex Method (Rough Cut)

No.	doc (mm)	Speed (m/mm)	Feed (mm/rev)	Time (min)	Cost (Rs)	Power (kW)	Force (N)	Temp. (°)	Roughness (μm)
1	2.00	118.32	0.750	2.87	27.54	3.90	710	496	9.30
2	2.10	118.32	0.749	2.88	27.58	4.00	736	498	9.40
3	2.20	113.13	0.738	2.94	27.78	3.90	758	490	10.63
4	2.30	113.13	0.738	2.95	27.81	4.10	783	493	10.15
5	2.40	113.13	0.738	2.96	27.84	4.20	809	495	10.26
6	2.50	113.13	0.738	2.97	27.88	4.30	834	497	10.36
7	2.60	113.13	0.738	2.98	27.91	4.50	859	499	10.46
8	2.70	117.50	0.656	3.10	27.35	4.30	809	497	8.86
9	2.80	117.50	0.656	3.11	28.39	4.40	836	499	8.94
10	2.90	110.22	0.692	3.11	28.34	4.50	892	493	10.49
11	3.00	108.67	0.660	3.15	28.62	4.40	885	487	10.30
12	3.10	111.74	0.640	3.18	28.72	4.50	855	491	9.66
13	3.20	109.12	0.602	3.27	29.12	4.30	869	482	9.49
14	3.30	111.28	0.603	3.31	29.09	4.50	889	488	9.30
15	3.40	126.25	0.510	3.39	29.07	4.50	794	498	6.53
16	3.50	117.50	0.546	3.44	29.53	4.60	859	492	7.86
17	3.60	113.13	0.510	3.58	30.04	4.30	839	479	7.83
18	3.70	113.13	0.510	3.60	29.97	4.50	852	489	7.38
19	3.80	113.13	0.510	3.61	30.12	4.40	873	482	7.94
20	3.90	113.13	0.510	3.62	30.15	4.50	890	483	7.99
21	4.00	126.25	0.473	3.69	30.41	4.80	850	499	6.31
22	4.10	122.86	0.497	3.64	30.23	5.00	900	500	6.95
23	4.20	123.60	0.485	3.69	30.40	5.00	900	500	6.76
24	4.30	124.08	0.474	3.74	30.50	5.00	900	500	6.62
25	4.40	124.10	0.463	3.79	30.76	5.00	900	499	6.48
26	4.50	124.45	0.452	3.84	30.96	5.00	900	498	6.34
27	4.60	124.53	0.442	3.89	31.12	5.00	900	497	6.23
28	4.70	124.66	0.433	3.94	31.31	5.00	900	496	6.12
29	4.80	124.84	0.423	4.00	31.49	5.00	900	495	6.01
30	4.90	124.96	0.414	4.05	31.61	5.00	900	494	5.91
31	5.00	124.10	0.406	4.10	31.86	5.00	900	494	5.80

TABLE 6.2
Optimization Results of Single-Pass Turning Using Nelder–Mead Simplex Method (Finish Cut)

No.	doc (mm)	Speed (m/min)	Feed (mm/rev)	Time (min)	Cost (Rs)	Power (kW)	Force (N)	Temp. (°)	Roughness (μm)
1	0.60	161.37	0.752	2.50	26.24	2.10	280	495	4.31
2	0.70	155.47	0.751	2.54	26.38	2.30	315	495	4.73
3	0.80	153.32	0.750	2.56	26.46	2.50	348	499	4.99
4	0.90	143.24	0.761	2.60	26.61	2.60	387	493	5.78
5	1.00	142.86	0.747	2.64	26.72	2.70	413	497	5.85
6	1.10	135.35	0.749	2.68	26.88	2.80	447	491	6.52
7	1.20	137.19	0.751	2.68	26.89	3.00	478	499	6.55
8	1.30	126.01	0.761	2.74	27.00	3.00	516	487	7.70
9	1.40	126.91	0.740	2.77	27.20	3.10	535	490	7.55
10	1.50	128.09	0.734	2.79	27.25	3.33	559	494	7.51

6.1.6 IMPLEMETATION OF GA

6.1.6.1 Binary Coding (V)

To solve this problem using GA, binary coding is chosen to represent the variables V and f. In the calculation here, 10 bits are chosen for V and 9 bits for f, making the total string length equal to 19.

No.	Code	Decode	V
1	0000000000	0	30
2	1111111111	1023	203.91
3	1001000110	582	128.94

6.1.6.2 Binary Coding (f)

No.	Code	Decode	F
1	000000000	0	0.254
2	111111111	511	0.765
3	1001000110	266	0.520

With the above coding, the following solution accuracy is obtained in the given interval:

	V (m/min)	f (mm/rev)
Accuracy	0.17	0.001
Interval	(30, 200)	(0.254, 0.765)

6.1.7 FITNESS FUNCTION

This is a constrained optimization problem, therefore penalty terms corresponding to the constraint violation are added to the objective function and a fitness function is obtained. Penalty terms are added only if the constraints are violated.

Fitness function (FFN) =

$$\text{Total production time} + \left[\frac{F - F_{max}}{F_{max}} + \frac{P - P_{max}}{P_{max}} + \frac{\theta - \theta_{max}}{\theta_{max}} + \frac{R_a - R_{a,max}}{R_{a,max}} \right] \times 10,000$$

6.1.8 REPRODUCTION

The rank selection method is used for reproduction. The linear ranking method proposed by Baker is as follows: each individual in the population is ranked in increasing order of fitness from 1 to N. The expected value of each individual i in the population at time t is given by

Expected value $(i,t) = \text{Min} + (\text{max} - \text{min}) (\text{rank}(i,t) - 1/ N{-}1)$

where $N = 20$. Minimum and maximum values for the above equation are obtained by performing a reproduction with the following set of values:

No.	Max	Min	P_{s1}	P_{s20}
1	1.1	0.9	0.045	0.055
2	1.5	0.5	0.025	0.075
3	1.6	0.4	0.020	0.080

where

P_{s1} = Probability of selecting the first rank
P_{s20} = Probability of selecting the 20th rank

From the above results, to have very low selection pressure for the first rank and high selection pressure for the twentieth rank, and to avoid quick convergence, maximum and minimum values are selected as 1.6 and 0.4, respectively. After calculating the expected value of each rank, reproduction is performed using Monte Carlo simulation by employing random numbers. The probability of selecting each rank is calculated and is given in Table 6.3. For example, generate random no. (between 0 to 1), if it is 0.330, select rank 10.

6.1.9 CROSSOVER

Flipping a coin with a probability 0.8 is simulated as follows: A three-digit decimal value between 0 and 1 is chosen at random. If the random number is smaller than 0.8, the outcome of the coin flip is true; otherwise, the outcome is false.

The next step is to find a cross-site at random. A crossover site is chosen by creating a random number between 1 and 9 for the first ten digits and 1 and 8

TABLE 6.3
Probability of Selection

Sol. No.	Expected Value	Probability of Selection	Cumulative Probability
1	$0.4 + 1.2 \times 0 = 0.4$	0.020	0.020
2	$0.4 + 1.2 \times 1/19 = 0.463$	0.023	0.043
3	$0.4 + 1.2 \times 2/19 = 0.526$	0.026	0.069
4	$0.4 + 1.2 \times 3/19 = 0.590$	0.030	0.098
5	$0.4 + 1.2 \times 4/19 = 0.652$	0.033	0.130
6	$0.4 + 1.2 \times 5/19 = 0.715$	0.036	0.165
7	$0.4 + 1.2 \times 6/19 = 0.779$	0.039	0.203
8	$0.4 + 1.2 \times 7/19 = 0.842$	0.042	0.244
9	$0.4 + 1.2 \times 8/19 = 0.905$	0.045	0.288
10	$0.4 + 1.2 \times 9/19 = 0.969$	0.048	0.335
11	$0.4 + 1.2 \times 10/19 = 1.03$	0.052	0.385
12	$0.4 + 1.2 \times 11/19 = 1.09$	0.055	0.438
13	$0.4 + 1.2 \times 12/19 = 1.16$	0.058	0.493
14	$0.4 + 1.2 \times 13/19 = 1.22$	0.061	0.551
15	$0.4 + 1.2 \times 14/19 = 1.28$	0.064	0.612
16	$0.4 + 1.2 \times 15/19 = 1.35$	0.068	0.676
17	$0.4 + 1.2 \times 16/19 = 1.41$	0.071	0.743
18	$0.4 + 1.2 \times 17/19 = 1.47$	0.074	0.814
19	$0.4 + 1.2 \times 18/19 = 1.54$	0.077	0.889
20	$0.4 + 1.2 \times 19/19 = 1.6$	0.080	1.000

for the other nine digits. For example, if the random number is 8, the strings are crossed at the site 8 and 2 new strings are created. After crossover, the new strings are placed in the intermediate population.

6.1.10 MUTATION

For bitwise mutation, a coin flip with a probability $P_m = 0.05$ is simulated using random numbers for every bit. If the outcome is true, the bit is reversed to 1 or 0, depending on the bit value.

6.1.10.1 GA Parameters and Result

Sample size = 20
Crossover probability (P_c) = 0.8
Mutation probability (P_m) = 0.05
Number of generations = 100

The result obtained by GA is given in Table 6.4 and the comparative data is given in Table 6.5, Table 6.6, and Table 6.7. The solution history is shown in Figure 6.2

TABLE 6.4
Optimization Results by GA for Turning Process

V (m/min)	f (mm/rev)	T_u (min)	C_T (U.S. Dollars)
114.49	0.665	3.13	28.46
P (kW)	F	(θ)	R_a
4.6	885.47	498.61	9.59

For 3 mm depth of cut.

TABLE 6.5
Comparison of Results — Turning Process

No.	Method	V	f	T_u
1	NMS	108.64	0.660	3.15
2	BSP	114.02	0.680	3.11
3	GA	114.49	0.665	3.13

For 3 mm depth of cut.

NMS – Nelder Mead simplex method

BSP – Boundary search procedure

TABLE 6.6
Results Obtained for Turning Process Optimization Using Conventional and Nonconventional Techniques

		BSP	NMS		GA		SA	
No.	doc	T_u	T_u	% dev	T_u	% dev	T_u	% dev
1	2.0	2.84	2.87	+1.0	2.85	+0.4	2.85	+0.4
2	2.5	2.93	2.97	+1.0	3.12	+1.0	2.93	0
3	3.0	3.11	3.15	+1.0	3.13	+1.0	3.15	+1.0
4	3.5	3.34	3.44	+3.0	3.46	+3.0	3.34	0
5	4.0	3.59	3.69	+3.0	3.51	+3.0	3.59	0
6	4.5	3.84	3.88	+1.0	3.96	+1.0	3.85	+0.3
7	5.0	4.10	4.23	+3.0	4.14	+3.0	4.12	+0.5

Total production time – T_u in min.

TABLE 6.7
Results Obtained for Turning Process Optimization Using Conventional and Nonconventional Techniques

No.	doc	BSP C_T	NMS C_T	% dev	GA C_T	% dev	SA C_T	% dev
1	2.0	27.45	27.54	+ 0.3	27.46	+ 0.04	27.49	+ .015
2	2.5	27.76	27.88	+ 0.43	28.43	+ 2.41	27.77	+ 0.04
3	3.0	28.37	28.62	+ 0.88	28.46	+ 0.32	28.54	+ 0.60
4	3.5	29.21	29.53	+ 1.1	29.62	+ 1.4	29.21	0
5	4.0	30.06	30.41	+ 1.16	29.79	+ 0.90	30.08	+ 0.07
6	4.5	30.94	30.80	+ 0.46	31.35	+ 1.33	30.99	+ 0.06
7	5.0	30.86	31.64	+ 0.70	31.64	+ 0.70	31.95	+ 0.28

Total production cost – C_T in Rs/piece.

6.2 OPTIMIZATION OF MULTI-PASS TURNING PROCESS

6.2.1 IMPLEMENTATION OF DYNAMIC PROGRAMMING TECHNIQUE

In multi-pass optimization problems, decisions must be made sequentially for the rough pass and finish pass. The problems in which the decisions are made sequentially are called sequential decision problems. Because these decisions are

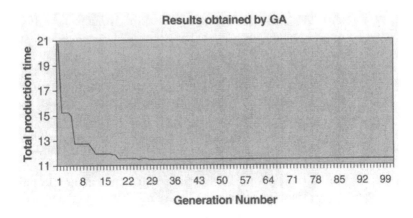

FIGURE 6.2 Results obtained by GA for turning process optimization.

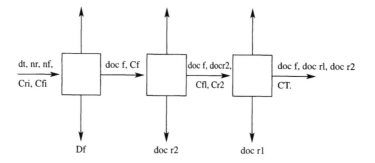

Dynamics programming model of the multi-pass turning process

Symbols used in the above model and dynamic programming flow chart

nf	-	number of finish passes
nf	-	number of finish passes
doc ri	-	depth of cut in the ith (rough)
doc fi	-	depth of cut in the ith (finish)
vri	-	cutting speed in the ith pass (rough)
vfi	-	cutting speed in the ith pass (finish)
fi	-	feed rate in the ith pass (rough)
fsi	-	feed rate in the ith pass (finish)
cri	-	cost for the ith pass (rough)
cri	-	cost for the ith pass (finish)

The results are given in the table.

FIGURE 6.3 Dynamic programming model of multi-pass turning process.

made at a number of stages, they are also referred to as multi-stage decision problems. Dynamic programming (DP) is a mathematical technique well-suited to the optimization of the above problem.

In this section, the procedure using DP for finding the optimal combination of depth of cut, optimum number of passes, and total minimum production cost is described. Let n be the maximum number of rough passes required. The number of passes needed to remove the total depth of the material will be $(n + 1)$, including the finish pass. The total minimum production cost is obtained by the summation of the cost of the finish pass and the cost of n rough passes. The dynamic programming model of this problem is given in Figure 6.3. The flowchart of the computation technique using dynamic programming is given in Figure 6.4. The optimization result is given in Table 6.8.

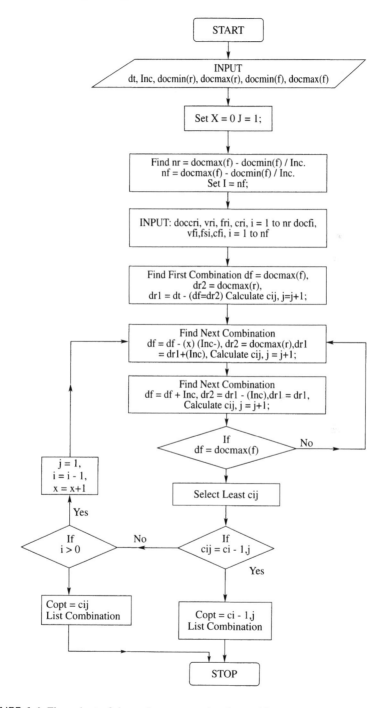

FIGURE 6.4 Flow chart of dynamic programming for multi-pass turning process.

TABLE 6.8
Optimization Results of Multipass Turning Operations
(Total Production Cost)

		Single Pass by Boundary Search Procedure					Single Pass by Nelder–Mead Simplex			
No.	d_t	d_f	D_{r1}	D_{r2}	N_{opt}	C_{opt}	d_f	D_{r1}	D_{r2}	N_{opt}	C_{opt}
1	5.0	1.5	3.5	—	2	56.30	1.5	3.5	—	2	56.95
2	5.5	1.5	4.0	—	2	57.15	1.5	4.2	—	2	57.55
3	6.0	1.5	4.5	—	2	58.06	1.5	4.5	—	2	58.33
4	6.5	1.5	5.0	—	2	58.96	1.5	5.0	—	2	59.54
5	7.0	1.5	1.5	4.0	3	84.25	1.5	2.6	2.9	3	83.55
6	7.5	1.5	2.4	3.6	3	84.17	1.5	2.9	3.1	3	84.36
7	8.0	1.5	2.9	3.6	3	84.57	1.5	2.6	3.9	3	85.31
8	8.5	1.5	2.7	4.3	3	85.59	1.5	2.6	4.4	3	86.11
9	9.0	1.5	2.7	4.8	3	86.40	1.5	3.1	4.4	3	86.92
10	9.5	1.5	3.1	4.9	3	87.25	1.5	3.3	4.7	3	87.85
11	10	1.5	3.6	4.9		88.08	1.5	4.0	4.5	3	88.47

d_t – Total depth of cut (mm).

d_f – Finish cut (mm).

d_r – Rough cut (mm).

N_{opt} – Optimum number of passes.

C_{opt} – Optimum cost (Rs/piece).

6.3 OPTIMIZATION OF FACE MILLING PROCESS

6.3.1 OBJECTIVE FUNCTION

Production rate is considered to be an objective function. The production rate can be expressed by the total processing time for one component (T_{ed}),

$$Q = 1/T_{ed}$$

where the processing time per component is determined as,

$$T_{ed} = T_m + T_{cm} + T_r$$

where T_m is the processing time, T_{cm} is the time for tool changing and setting up and T_r is the time for manual operations. All times are given in minutes.

For the face milling process, the production time is defined as

$$T_m = \pi DL/(1000 ZVf)$$

where

$$L = L_b + L_n + L_a$$

and where D is the diameter of the cutter, Z is the number of edges, V is the cutting speed, f is the feed rate, L_b is the length of movement before cutting, L_a is the length of movement after cutting, and L_n is the acting surface length.

6.3.2 MACHINING VARIABLES

In this work, controllable variables such as cutting speed, feed rate, and depth of cut are considered. Depth of cut can often be determined by the size of the work material and product; hence, it is assumed constant. The machining variables are discrete in the case of a conventional machine tool and continuous in the case of a CNC machine tool. Because of the predominance of CNC machining, the variables are treated as continuous.

6.3.3 MACHINING CONSTRAINTS

The following constraints are considered as reflecting the kinematic and dynamic features of a face milling operation:

Bounds of cutting speed:

$$V_{min} \leq V \leq V_{max}$$

Bounds of feed rate:

$$f_{min} \leq f \leq f_{max}$$

Minimum and maximum value of tool life:

$$T_{min} \leq T \leq T_{max}$$

Power of the main drive:

$$CFt \cdot V^{NFt+1} \, f^{Yt} \, d^{XFt}/60{,}000 \leq P_{m,max} \ (kW)$$

Power of the feed drive:

$$Cfp \, V^{NFtp+1} \, f^{Yfp} \, d^{XFP}/60{,}000 \leq P_{f,max} \ (kW)$$

Maximum allowable tangential cutting force
$$CFt \, V^{NFt} \, f^{Yt} \, d^{XFt} / 60000 \leq F_{t,max}$$
Maximum allowable surface roughness:

$$CRa \, V^{NRa} \, f^{YRa} \leq R_{a,max} \ (\mu m)$$

The values of parameters C_{ft}, N_{ft}, Y_{ft}, C_{fp}, N_{fp}, Y_{fp}, C_{ra}, N_{ra}, and Y_{ra} that are present in the constraints depend upon the workpiece and tool material. The parameter d is the depth of cut.

6.3.4 DATA OF THE PROBLEM

PROCESS VARIABLES	SYMBOL	VALUE
Speed of the rapid traverse (m/min):	V_b	2500
Dimeter of the mill (mm):	D	160
Number of mounting faces:	E_t	10
Number of cutting edges in a tip:	E_p	4
	C_t	860
	X_t	0.87
	Y_t	1.17
	N_t	2.06
Tangential force parameters:	Q_t	0.80
	C_{Ft}	4230
	X_{Ft}	1.00
	Y_{ft}	0.75
Radial force parameters:	N_{Ft}	0.05
Radial force parameters:	C_{Fp}	10,404
	F_p	0.85
	Y_{fp}	1.04
	N_{Fp}	0.33
Surface roughness parameters:	C_{ra}	24
	Y_{ra}	0.73
	N_{ra}	0.33
Acting surface length:	L_n	200
Width of cut (mm):	B	32
Length of movement before cutting (mm):	L_b	10
Length of movement after cutting (mm):	L_a	90
Cost of tips (Rs):	L_a	90
Cost of body (Rs):	C_p	50
Time of manual operations (min):	C_i	2000
Time of manual operations (min):	T_r	0.12
Time for tool changing and setting (min):	T_{cm}	1
Minimum allowable cutting speed (rev/min):	V_{min}	60
Minimum allowable cutting speed (rev/min):	V_{max}	125
Minimum allowable depth of cut (mm):	d_{min}	1
Minimum allowable feed rate (mm/tooth):	f_{min}	0.1
Minimum allowable feed rate (mm/tooth):	f_{max}	0.2
Minimum allowed power of feed drive (kW):	$P_{f,max}$	2.5
Minimum allowed power of main drive (kW):	$P_{m,max}$	7.5
Maximum allowable surface roughness (μm):	$R_{a,max}$	2.0

6.3.5 IMPLEMENTATION OF GA FOR FACE MILLING PROCESS OPTIMIZATION

6.3.5.1 Binary Coding

For this problem, binary coding is chosen to represent the variables V and f. In the calculation here, 9 bits are chosen for V and 9 bits for f, making the total string length 18, which is given below.

(a) Coding (for V):

No.	Code	Decode	V (rev/min)
1	0000000000	0	63
2	1111111111	511	125.00
3	111001000	456	117.72

(b) Coding (for f):

No.	Code	Decode	f (mm/tooth)
1	000000000	0	0.1000
2	111111111	511	0.2000
3	001110011	115	0.1230

With this coding, the following solution accuracy is obtained for the given interval:

	V (m/min)	f (mm/tooth)
Accuracy	0.12	0.0002
Interval	(63, 126)	(0.1, 0.2)

6.3.6 FITNESS FUNCTION

This is a constrained optimization problem. Penalty terms corresponding to the constraint violations are added to the objective function and a fitness function is obtained. Penalty terms are added only if the constraints are violated.

Fitness function (FFN) = Production rate −

$$\left[\frac{F_t - F_{t,max}}{F_{t,max}} + \frac{P_m - P_{m,max}}{P_{m,max}} + \frac{P_f - P_{f,max}}{P_{f,max}} + \frac{T - T_{max}}{T_{max}} + \frac{R_a - R_{a,max}}{R_{a,max}} \right] \times 100$$

TABLE 6.9
Comparison of Results Obtained for Face Milling Process Optimization

No.	Method	V (rev/min)	F (mm/tooth)	Production rate (min/piece)
1	Linear Programming Technique	62.99	0.158	0.380
2	GA	99.12	0.166	0.938

6.3.7 GENETIC OPERATIONS

The genetic operations such as reproduction, crossover, and mutation are performed for this problem similar to the turning process problem.

6.3.8 OPTIMIZATION RESULTS

Results obtained for the face milling process optimization are compared with the linear programming method and given in Table 6.9.

6.4 SURFACE GRINDING PROCESS OPTIMIZATION

6.4.1 NOMENCLATURE

G Grinding ratio

k_a Constant dependent upon coolant and wheel grind type

k_u Wear constant (mm^1)

K_c Cutting stiffness (N/mm)

K_m Static machine stiffness (N/mm)

K_s Wheel-wear stiffness (N/mm)

L Lead of dressing (mm/rev)

L_e Empty length of grinding (mm)

L_w Length of workpiece (mm)

M_c Cost per hour, labor and administration ($/h)

N_d Total number pieces to be ground between two dressings

N_t Batch size of workpieces

N_{td} Total number of workpieces to be ground during the life of dresser

p Number of workpieces loaded on table

R_a Surface finish (mm)

R_a^* Surface finish limitation during rough grinding (mm)

R_c Workpiece hardness (Rockwell hardness number)

R_{em} Dynamic machine characteristics

S_d Distance of wheel idling (mm)

S_p Number of sparkout grinding (pass)

t_{sh} Time of adjusting machine tool (min)

t_i Time of loading and unloading workpiece (min)

T_{ave} Average chip thickness during grinding (mm)
U Specific grinding energy (J/mm³)
U^* Critical specific grinding energy (J/mm³)
V_r Speed of wheel idling (mm/min)
V_s Wheel speed (m/min)
V_w Workpiece speed (m/min)
VOL Wheel bond percentage
WRP Workpiece removal parameter (mm³/min-N)
WRP^* Workpiece removal parameter limitation (mm³/min-N)
WWP Wheel wear parameter (mm³/min-N)
W_1, W_2 Weighting factors, $0 \le W_i \le 1$ ($W_1 + W_2 = 1$)

6.4.2 DETERMINATION OF SUBOBJECTIVES AND VARIABLES FOR OPTIMIZATION

The aim of carrying out grinding operations is to obtain the finished product with a minimum production cost, maximum production rate, and the finest possible surface finish. Therefore the production cost and production rate were chosen as subobjectives for the surface grinding process. The resultant objective function of the process is a weighted combination of the two such objectives through a weighted approach. Since numerous process variables are involved in grinding, especially where changes are extremely sensitive to the final performance of the parts, optimizing every variable is difficult and complex. Fortunately, among the numerous process variables, some are determined by operators and some are of greater importance than others. This usually provides the guide to the selection of the variables, which are considered the optimization parameters. In this example, four variables — namely, wheel speed, workpiece speed, depth of dressing and lead of dressing — are considered the optimization variables.

6.4.3 RELATIONSHIPS BETWEEN TWO SUBOBJECTIVES AND FOUR OPTIMIZATION VARIABLES

6.4.3.1 Production Cost

In the surface grinding process, the production costs include three elements: the cost directly related to the grinding of the part, the cost of nonproductive time and the cost of material consumption. The total production cost during the grinding process, C_T, considering the various elements mentioned above, is shown in Equation 6.1:

$$CT = \frac{M_C}{60p}\left(\frac{L_w + L_e}{V_w 1000}\right)\left(\frac{b_w + b_e}{f_b}\right)\left(\frac{a_w}{a_p} + Sp + \frac{a_w b_w L_w}{\pi \, D_c b_s a_p G}\right) + \frac{M_c}{60p}\left(\frac{S_d}{V_t} + t_1\right)$$

$$\frac{M_c t_{ch}}{60 N_t} + \frac{M_c \cdot 1\pi \, b_s D_e}{60 p N_d LV_s 1000} + C_s\left(\frac{a_w b_w L_w}{pG} + \frac{\pi \, docb_s D_e}{pN_d}\right) + \frac{C_d}{pN_{td}}$$

(6.1)

6.4.4 Production Rate

The production rate is represented by the work piece removal parameter, *WRP*. The *WRP* is directly related to the grinding conditions and details of wheel dressing preceding the grinding operations.

$$WRP = 94.4 \frac{\left(1 + \dfrac{2DOC}{3L}\right) L^{11/19} \left(\dfrac{V_w}{V_s}\right)^{3/19} V_s}{D_c^{43/403} VOL^{0.47} d_g^{5/38} R_c^{27/19}} \tag{6.2}$$

where $VOL = 1.33X + 2.2S - 8$, and where the values of $X = 0, 1, 2, 3...$, for wheel hardness of H, I, J, K..., respectively, and S is the wheel structure number, i.e., 4, 5, 6....

6.4.5 Constraints

A more complete solution to the grinding problem takes into account several realistic constraints of the actual operations. The constraints can be divided into process constraints and variable constraints. The process constraints considered in the present work are thermal damage, a wheel wear parameter, machine tool stiffness, and surface finish. The variable constraints are the upper and lower limits of the grinding conditions.

6.4.6 Thermal Damage Constraints

Because grinding processes require an extremely high input of energy per unit volume of material removed, and convert almost all the energy into heat that is concentrated within the grinding zone, the high thermal energy can cause thermal damage to the workpiece. One of the most common types of thermal damage is workpiece burn, which directly limits the production rate. On the basis of heat transfer analysis and experimental measurements, burning has been shown to occur when a critical grinding zone temperature is reached. This temperature is directly related to the specific energy, which consists of chip formation energy, ploughing energy, and sliding energy. Combining the relationships, the specific grinding energy, U, is given in terms of the operating parameters by Equation 6.4.

$$U = 13.8 \frac{9.64 \times 10 - 4V_s}{apVw} + \left(6.9 \times 10 - 3 + \frac{2102.4Vw}{DeVs}\right) \times \frac{VsDe^{1/2}}{Vwap^{1/2}} \tag{6.3}$$

The corresponding critical specific grinding energy U^* at which burning starts can be expressed in terms of the operating parameters as

$$U^* = 6.2 + 1.76\left(\frac{De^{1/4}}{ap^{3/4}Vw^{1/2}}\right) \tag{6.4}$$

In practice, the specific energy must not exceed the critical specific energy $U*$; otherwise, workpiece burn occurs. According to the relationship between grinding parameters and the specific energy (Equation 6.4), the thermal damage constraint can be specified as

$$U \leq U*$$

6.4.7 Wheel Wear Parameter Constraint

Another constraint is the wheel wear parameter WWP, which is related to the grinding conditions and the details of wheel dressing preceding the grinding operations.

$$WWP = \left(\frac{\left(1 + doc/L\right) L^{27/19} \left(V_s / V_w^{3/19} V_w\right)}{1 + 2doc/3L} \right) \tag{6.5}$$

The wear of the grinding wheel (grinding ratio) is usually expressed as the volumetric loss of material, WWP, and is determined by the typical wheel wear behavior given by a plot of WWP versus the accumulated workpiece removal (*WRP*). According to Equation 6.2 and Equation 6.7, the wheel wear constraint can be obtained as follows:

$$WRP/WWP \geq G$$

6.4.8 Machine Tool Stiffness Constraint

In grinding, chatter results in undulation roughness on the grinding wheel or workpiece surface and is highly undesirable. A reduction of the workpiece removal rate is usually required to eliminate grinding chatter. In addition, wheel surface unevenness necessitates frequent wheel redressing. Thus, chatter results in a worsening of surface quality and lowers the machining production rate. Avoiding chatter is therefore a significant constraint in the selection of the operating parameters.

The relationship between grinding stiffness K_c, wheel wear stiffness K_s and operating parameters during grinding is expressed as follows:

$$K_c = \frac{1000 V_w f_b}{WRP} = \frac{1000 D_c^{43/304} VOL^{0.47} d_g^{5/38} R_c^{27/19} f_b \left(V_w / V_s\right)^{16/19}}{94.4 \left(1 + 2doc / 3L\right) L^{11/19}} \tag{6.6}$$

$$K_s = \frac{1000 V_s f_b}{WRP} = \frac{1000 D_c^{1.2/vol - 43/304} VOL^{0.38} f_b}{K_a a_p d_g^{5/38} R_c^{27/19}} \times \frac{\left(1 + 2doc / 3L\right) \left(V_s / V_w\right)^{16/19}}{\left(1 + doc / L\right)^{27/19}} \tag{6.7}$$

Grinding stiffness and wheel wear stiffness during grinding as well as the static machine stiffness must satisfy the following constraint to avoid excessive chatter during grinding:

$\quad$ MSC $\geq$ |Rem|/Km

$\quad$ Where

$$MSC = \frac{1}{2K_C}\left(1+\frac{Vw}{V_sG}\right)\frac{1}{K_S} \tag{6.8}$$

6.4.9 SURFACE FINISH CONSTRAINT

The surface finish, R_a, of a workpiece is usually specified within a certain value R_a^*. The operation parameters and wheel dressing parameters strongly influence the surface finish.

$$T_{ave} = 12.5\times10^3\frac{d_g^{16/27}a_p^{19/27}}{D_c^{8/27}}\left(1+\frac{doc}{L}\right)L^{16/25}\left(\frac{V_w}{V_s}\right)^{16/27} \tag{6.9}$$

$$R_a = \begin{cases} 0.4587T_{ave}^{0.30} & \text{for } O<T_{ave}<0.254 \\ 0.7866T_{ave}^{0.72} & \text{for } 0.254<T_{ave}<2.54 \end{cases} \tag{6.10}$$

6.4.10 RESULTANT OBJECTIVE FUNCTION MODEL

Through the analysis discussed above, the optimization problem for the surface grinding process can be formulated as a multi-objective, multivariable, nonlinear optimization problem with multiple constraints. To overcome the large differences in numerical values between the subobjectives, normalization of each subobjective is introduced. The resultant weighted objective function to be minimized here is:

$$COF\left(V_s,V_w,doc,L\right) = W_1\frac{CT}{CT^*} - W_2\frac{WRP}{WRP^*} \tag{6.11}$$

$\quad$ Subject to:

$$U \leq U^*$$

$$WRP/WWP \geq G$$

$$MSC \geq |R_{em}|/K_m$$

$$R_a \leq R_a^* \text{ (for rough grinding)}$$

$$WRP \geq WRP^* \text{ (for finish grinding)}$$

6.4.11 Data of Problem

DESCRIPTION	SYMBOL	VALUE
Number of workpieces loaded on table:	p	1
Length of the workpiece (mm):	L_w	300
Empty length of grinding (mm):	L_e	150
Width of workpiece (mm):	b_w	60
Empty width of grinding (mm):	b_e	25
Total thickness of cut (mm):	a_w	0.1
Down feed of grinding (mm/pass):	a_p	0.0505
Number of sparkout grinding (pass):	S_p	2
Distance of wheel idling (mm):	S_d	100
Speed of wheel idling (mm/min):	V_r	254
Time of loading and unloading workpiece (min):	t_l	5
Time of adjusting machine tool (min):	t_{ch}	30
Total number of workpieces ground between two dressings:	N_d	20
Batch size of workpiece:	N_t	12
Total number of workpieces ground during life of dresser:	N_{td}	2000
Cost of wheel per mm³ (Rs/mm³):	C_s	0.120
Workpiece hardness (Rockwell):	R_c	58
Surface finish limitation, rough (μm):	$R_a{}^*$	1.8
Workpiece removal parameter limitation:	WRP^*	20
Static machine stiffness (N/mm):	K_m	100,000
Dynamic machine characteristics:	R_{em}	1
Initial wear flat area percentage:	A_0	0
Wear constant (mm¹):	k_u	3.937×10^7
Constant dependent on coolant and grain:	k_a	0.0869

6.4.12 Implementation of GA for Four Variable Problems

6.4.12.1 Binary Coding

To solve this problem using GA, binary coding is chosen to represent the variables V_s, V_w, doc and L. In the calculation here, 10 bits are chosen for V_s and 7 bits each for the other three variables, making the total string length equal to 31.

	V_s	V_w	doc	L
Binary code	100100010	1111110	0001011	0001110
Decoded valve	1610	22.6	0.021	0.024

With this coding, the following solution accuracy is obtained in the given interval:

	V_s (m/min)	V_2 (m/min)	doc (mm)	l (mm/rev)
Accuracy	1.0	0.1	0.001	0.001
Interval	(1000, 2023)	(10, 22.7)	(0.01, 0.137)	(0.01, 0.137)

6.4.12.2 Fitness Function

This is a constrained optimization problem. Penalty terms corresponding to the constraint violations are added to the objective function and a fitness function is obtained. Penalty terms are added only if the constraints are violated.

$$Fitness\ Function,\ FFN = NOF - \frac{\left(U - U^*\right)}{U} - \left(\frac{G - \left(\dfrac{WRP}{WWP}\right)}{G}\right) - \left(1 - MSC \times 10^5\right)$$
$$- \left(\frac{R_a - R_a^{\ *}}{R_a^{\ *}}\right)$$

$$(6.12)$$

where *MSC* is the machine tool stiffness constraint.

6.4.13 REPRODUCTION

The rank selection method is used for reproduction.

6.4.14 CROSSOVER

Similar to crossover performed for turning optimization, multipoint crossover is used for the grinding. A crossover site is chosen by creating a random number between 1 and 9 for the first ten digits and between 1 and 20 for the remaining balance of 21 digits. For example, the random number is 8, the strings are crossed at site 8 and two new child strings are created. After crossover, the child strings are placed in the intermediate population.

6.4.15 MUTATION

Bitwise mutation is performed, using the procedure for turning process optimization.
Results obtained by GA are given in Table 6.10 and Table 6.11.

TABLE 6.10
Comparison of Results (Rough Grinding — Four Variables)

No.	Method	V_s (n/min)	V (m/min)	doc (mm)	L (mm/rev)	C_T (Rs/piece)	WRP (mm³/min-N)	R_a (μm)	COF
1	QP	2000	19.96	0.055	0.044	6.2	17.47	1.74	0.127
2	GA	2018	15.60	0.070	0.028	6.26	18.50	1.79	0.149

QP — Quadratic Programming.

GA — Genetic Algorithm.

6.4.16 IMPLEMENTATION FOR TEN-VARIABLE SURFACE GRINDING OPTIMIZATION

6.4.16.1 Optimization Variables

Since numerous process variables are involved in grinding, especially where changes are extremely sensitive to the final performance of the parts, optimizing every variable is required. Unfortunately, among the numerous process variables, only four variables were considered so far because of the complexity of the solution. In addition, six more variables can be included (total of ten variables):

Wheel speed (V_s)
Workpiece speed (V_w)
Depth of dressing (doc)
Lead of dressing (L)
Cross feed rate (f_b)
Wheel diameter (D_e)
Wheel width (D_b)
Grain size (d_g)
Wheel bond percentage (VOL)
Grinding ratio (G)

TABLE 6.11
Comparison of Results (Finish Grinding — Four Variables)

No.	Method	V_s (m/min)	V_w (m/min)	doc (mm)	L (mm/rev)	C_T (Rs/piece)	WRP (mm³/min-N)	R_a (μm)	COF
1	QP	2000	19.99	0.052	0.091	7.7	20.00	0.83	0.554
2	GA	1986	21.40	0.24	0.136	6.6	20.08	0.83	0.521

QP — Quadratic programming.

GA — Genetic algorithm.

TABLE 6.12
Optimization Results by GA for
Surface Grinding (Ten Variables)

V_s	V_w	doc	L	f_b
2000	16.50	0.065	0.045	2.06

D_e	b_s	G	VOL	d_g
3.60	24.40	61.0	6.80	0.275

C_T	WRP	COF	NOF	FFN
243.20	19.01	−0.172	1.207	1.207

COF — Combined objective function.

NOF — New objective function.

FFN — Fitness function.

6.4.17 SPECIAL CODING

To solve this problem using GA, a special type of coding system is chosen to represent the variables V_s, V_w, doc, L, f_b, D_e, D_b, d_g, VOL, and G. The coding consists of 19 digits, where the first 10 digits are binary numbers (0 or 1) and the next 9 digits are numbers ranging from 0 to 9.

(e.g) Coding: 111110100 2 7 3 7 9 2 5 2 4
↓ ↓ ↓ ↓ ↓ ↓ ↓ ↓ ↓ ↓
V_s V_w d_{oc} L f_b D_e b_s G VOL d_g

Results obtained by GA are given in Table 6.12. The comparative analysis is given in Table 6.13.

TABLE 6.13
Comparison of Results Obtained for Surface Grinding Process
(Four and Ten Variables)

No.	Method	Variables	V_s (m/min)	V_w (m/min)	doc (mm)	L (mm/rev)	C_T (Rs/pc)	WRP (mm³/min-N)	COF
1	QP	Four	2000	19.96	0.55	0.044	6.20	17.47	0.127
2	GP	Four	2018	15.60	0.070	0.028	6.26	18.50	0.149
3	GA	Ten	2000	16.50	0.065	0.045	6.08	19.01	0.172

6.5 OPTIMIZATION OF MACHINING PARAMETERS FOR MULTI-TOOL MILLING OPERATIONS USING TABU SEARCH

6.5.1 Nomenclature

a, a_{rad}	Axial depth of cut, radial depth of cut (mm)
C	Constant in cutting speed equation
ca	Clearance angle of the tool (°)
C_i $(i = 1\text{–}8)$	Coefficients carrying constant values
C_i	Labor cost, co-overhead cost
C_m, C_{mat}	Machining cost, cost of raw material per part, cost of a cutting tool ($)
C_u	Unit cost ($)
d	Cutter diameter (mm)
e	Machine tool efficiency factor
F	Feed rate (mm/min)
f	Feed rate (mm/tooth)
G, g	Slenderness ratio, exponent of slenderness ratio
K	Distance to be traveled by the tool to perform the operation (mm)
K_i $(i = 1\text{–}3)$	Coefficients carrying constant values
K_p	Power constant depending on the work piece material
la	Lead (corner) angle of the tool (°)
m	Number of machining operations required to produce the product
N	Spindle speed (rev/min)
n	Tool life exponent
P	Required power for the operation (kW)
P_m	Motor power (kW)
P_r	Total profit rate ($/min)
Q	Contact proportion of cutting edge with work piece per revolution
S_p	Sale price of the product ($)
T	Tool life (min)
T_u	Unit time (min)
t_m, t_s, t_{tc}	Machining time, setup time, tool changing time (min)
V	Cutting speed (m/min)
w	Exponent of chip cross-sectional area
W	Tool wear factor
Z	Number of cutting teeth of the tool

6.5.2 Unit Cost

The unit cost based on cutting speed and feed rate is:

$$C_u = C_1 + \sum_{i=1}^{m} C_{2i} V_i^{-1} f_i^{-1} + \sum_{i=1}^{m} C_{3i} V_i^{(1/n)-1} f_i^{[(w+g)/n]-1} + \sum_{i=1}^{m} C_{4i}$$

where
$$C_1 = C_{mat} + (C_1 + C_0)t_s$$
$$C_{2i} = (C_1 + C_0)K_{1i}$$
$$C_{3i} = c_{ti} K_{3i}$$
$$C_{4i} = (C_1 + C_0)t_{tci}$$
$$K_1 = d\ K/1000z$$
$$K_2 = 60\ Q^1\ C^{1/n}\ 5^{g/n}\ a^{(gw)/n}$$
$$K_3 = K_1/K_2$$

C_1 represents material and setup costs, while C_2 represents the labour and overhead cost and C_3 represents the tooling costcost. C_4 represents tool changing cost.

6.5.3 Unit Time

The unit time is the sum of setup, machining, and tool changing times.

$$T_u = ts + \sum_{i=1}^{m} K_{1i}V_i^{-1}f_i^{-1} + \sum_{i=1}^{m} t_{tci}$$

Profit Rate
The total profit rate in machining can be determined by

$$P_r = \frac{S_p - C_u}{T_u}$$

6.5.4 Constraints

6.5.4.1 Power

The machining power is 0.8

$$P = \frac{0.78K_p\ Wza_{rad}a}{60\pi de} Vf^{0.8}$$

Required machining power for the operation P must not exceed the available motor power, P_m. Therefore, the power constraint can be written as

$$P = \frac{0.78K_p\ Wza_{rad}a}{60\pi deP_m} Vf^{0.8} \le 1$$

6.5.4.2 Surface Finish

The surface finish for face milling operation can be represented by

$$\frac{318[\tan(la) + \cot(ca)]^{-1}}{R_{a(at)}} f \le 1$$

TABLE 6.14
Random Initial String Solution

Str_vI Speed	Str_fl Feed	Str_v2 Speed	Str_a Feed	Str_v3 Speed	Str J3 Feed
1011100110	0001000010	0110001000	1111001101	0010111011	M0001111

Str_v4 Speed	Str f4 Feed	Str v5 Speed	Str_fS Feed
0110100100	0110010110	1100101100	0011011110

"TOTAL MAXIMUM PRODUCTION RATE"

vI	f1	v2	f2	v3	f3	v4	f4	v5	f5
103.519	0073	51.496	0.478	45.484	0.169	38.211	0.229	45.875	0.148

Cu	Tu	Pr
11.841	5.706	2.306

The surface finish for end milling operation is
$(318 \ (4d)^{-1} \ f^2)/R_a(at) \ \Sigma \ 1$

$$\frac{318(4d)^{-1}}{R_{a(at)}} f_2 \leq 1$$

Tabu search output is given in Tables 6.14–6.20.

TABLE 6.15
Iteration I

Iter	Bit	Str_vI	Str_f1	Str_v2	Str_f2
1	1	0011100110	1001000010	1110001000	0111001101
1	2	1111100110	0101000010	0010001000	1011001101
1	3	1001100110	0011000010	0100001000	1101001101
1	4	1010100110	0000000010	0111001000	1110001101
1	5	1011000110	0001100010	0110101000	1111101101
1	6	1011110110	0001010010	0110011000	1111011101
1	7	1011101110	0001001010	0110000000	1111000101
1	8	1011100010	0001000110	0110001100	1111001001

Str_v3	Str_f3	Str_v4	Str_f4	Str_v5	Str_f5
1010111011	11.00001111	1110100100	1110010110	0100101100	1011011110
0110111011	0000001111	0010100100	0010010110	1000101100	0111011110
0000111011	0110001111	0100100100	0100010110	1110101100	000101111 0
0011111011	0101001111	0111100100	0111010110	1101101100	0010011110
0010011011	0100101111	0110000100	0110110110	1100001100	0011111110
0010101011	0100011111	0110110100	0110000110	1100111100	0011001110
0010110011	0100000111	0110101100	0110011110	1100100100	0011010110
0010111111	0100001011	0110100000	0110010010	1100101000	0011011010

TABLE 6.16
Calculation of Profit Rate for First Iteration

v1	f1	v2	f2	v3	f3	v4	f4	v5	f5
73.490	0.248	66.510	0.253	60.499	0.394	48.221	0.454	35.865	0.373
118.534	0.160	43.988	0.365	52.991	0.057	33.206	0.116	40.870	0.260
96.012	0.116	47.742	0.422	41.730	0.226	35.709	0.172	48.377	0.091
99.765	0.051	53.372	0.450	47.361	0.197	39.462	0.257	47.126	0.120
101.642	0.084	52.434	0.492	44.545	0.183	37.586	0.243	45.249	0.162
104.457	0.078	51.965	0.485	45.015	0.176	38.524	0.222	46.188	0.141
103.988	0.075	51.261	0.474	45.249	0.166	38.368	0.232	45.718	0.144
103.284	0.074	51.613	0.476	45.601	0.167	38.133	0.227	45.797	0.146

Cu	T_u	P_r
11.816	4.410	2.990
11.459	5.495	2.464
10.793	5.192	2.736
13.211	6.406	1.840
11.459	5.488	2.467
11.616	5.575	2.401
11.742	5.658	2.343

TABLE 6.17
The Best (Tabu) Solution for First Iteration

str_v1	str_f1	str_v2	str_f2	str_v3	str_f3
0011100110	1001000010	1110001000	0111001101	1010111011	1100001111

str_v4	str_f4	str_v5	str_f5
1110100100	1110010110	0100101100	1011011110

$C_u = 11.816$ $T_u = 4.410$ $P_r = 2.990$

(Bit 1 until iteration 9)

TABLE 6.18
Iteration 2: Maximum Profit Rate Obtained in 2 Bit

Iter	Bit	Str_vl	Str_f1	Str_v2	Str_f2
2	9	0011100100	1001000000	1110001010	0111001111
2	10	0011100111	1001000011	1110001001	0111001100
2	1	Tabu			
2	2	0111100110	1101000010	1010001000	0011001101
2	3	0001100110	1011000010	1100001000	0101001101
2	4	0010100110	1000000010	1111001000	0110001101
2	5	0011000110	1001100010	1110101000	0111101101
2	6	001111 0110	1001010010	1110011000	0111011101

Str_v3	Str_f3	Str_v4	Str_f4	Str_v5	St_f5
1010111001	1100001101	1110100110	1110010100	0100101110	1011011100
1010111010	1100001110	1110100101	1110010111	0100101101	1011011111
Tabu					
1110111011	1000001111	1010100100	1010010110	0000101100	1111011110
1000111011	1110001111	1100100100	1100010110	0110101100	1001011110
1011111011	1101001111	1111100100	1111010110		
1010011011	1100101111	1110000100	1110110110		
1010101011	1100011111	1110110100	1110000110		

TABLE 6.19
Calculation of Profit Rate for Second Iteration

"TOTAL MAXIMUM PROFIT RATE"									
v1	f1	v2	f2	v3	f3	v4	f4	v5	f5
73.372	0.247	66.569	0.254	60.440	0.394	48.260	0.453	35.904	0.372
73.548	0.248	66.540	0.252	60.469	0.394	48.240	0.454	35.885	0.373
tabu									
88.504	0.335	59.003	0.140	68.006	0.282	43.216	0.341	30.860	0.485
65.982	0.292	62.757	0.196	56.745	0.451	45.718	0.398	38.368	0.317
69.736	0.226	68.387	0.225	62.375	0.423	49.472	0.482	37.116	0.345
71.613	0.259	67.449	0.267	59.560	0.409	47.595	0.468	35.239	0.387
74.428	0.253	66.979	0.260	60.029	0.401	48.534	0.447	36.178	0.366

C_a	T_u	P_r
11.801	4.413	2.991
11.863	4.408	2.994
Tabu		
11.546	4.240	3.173
11.324	4.389	3.116

TABLE 6.20
The Best (Tabu) Solution for Second Iteration

str_vl	str_fl	str_v2	str_f2	str_v3	str_f3
0111100110	1101000010	1010001000	0011001101	1110111011	1000001111

str_v4	str_f4	str_v5	str_f5	p_r	
1010100100	1010010110	0000101100	1111011110	3.173	

C_u =11.546; T_u = 4.240; P_r = 3.173.

(Bit 2 until iteration 10)

REFERENCES

Agapiou, J.S., The optimization of machining operations based on a combined criterion, part 1: the use of combined objectives in single pass operations, *Transactions of ASME, Journal of Engineering for Industry,* 114, 500–507, 1992.

Agapiou, J.S., The optimization of machining operations based on a combined criterion, part 2: multipass operations, *ASME Journal of Engineering for Industry,* 114, 508–513, 1992.

Amitay, G., Adaptive control optimization of grinding, *ASME Journal of Engineering for Industry,* v103, 103–108, 1981.

Asokan, P., et al., Machining parameters optimization for turning cylindrical stock into a continuous finished profile using genetic algorithm (GA) and simulated annealing (SA), *International Journal of Advanced Manufacturing Technology,* 21(1), 1–9, 2003.

Gupta, R. et al., Determination of optimal subdivision of depth of cut in multipass turning with constraints, *International Journal of Production Research,* 33(9), 2555–2565, 1994.

Jha, N.K. and Rao, V.R.K., Opimization of milling process through geometric programming, *Proceedings of the Third International Conference on Production Engineering,* Kyoto, Japan, 1977.

Malkin, S., Practical approaches to grinding optimization, in M.C. Shaw, Grinding Symposium, *ASME Winter Annual Meeting,* Miami Beach, FL, 289–299, 1985.

Saravanan, R. et al., Comparative analysis of conventional and non-conventional optimization techniques for CNC turning process, *International Journal of Advanced Manufacturing Technology,* 17(7), 471–476, 2001.

Saravanan, R., et al., Genetic algorithm (GA) for multivariable surface grinding process optimization using a multi-objective function model, *International Journal of Advanced Manufacturing Technology,* 17(5), 330–338, 2001.

Saravanan, R., Vengadesan, S., and Sachithanandam, M., Selection of operating parameters in surface grinding process using genetic algorithm (GA), *Proceedings of the 18th All-India Manufacturing Technology, Design and Research Conference,* Kharagpur, India, 167–171, 1998.

Shin, Y.C. et al., Optimization of machining conditions with practical constraints, *International Journal of Production Research*, 30, 2907–2919, 1992.

Vijayakumar, K. et al., Optimization of multipass turning operations using ant colony system, *International Journal of Advanced Manufacturing Technology,* 43(15), 1633–1639, 2003.

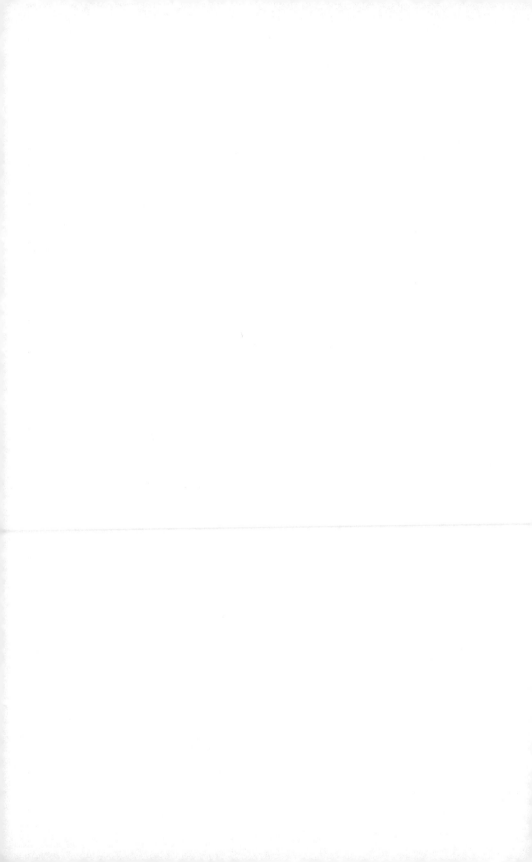

7 Integrated Product Development and Optimization

7.1 INTRODUCTION

The convention of designing products is characterized as an iterative procedure that consists of six identifiable steps or phases. Figure 7.1 illustrates the six basic steps in the design process, indicating its iterative nature.

Recognition of need: This step involves the realization by someone that a problem exists for which some corrective action should be taken. This might be the identification of some defect in a current machine design by an engineer or the perception of a new product marketing opportunity by a salesperson.

Definition of problem: This step involves a thorough specification of the item to be designed. This specification includes physical and functional characteristics, cost, quality, and operating performance.

Synthesis and analysis: These two steps are closely related and highly iterative in their processes. A certain component or subsystem of the overall system is conceptualized by the designer, subjected to analysis, improved through this analysis procedure, and redesigned. The process is repeated until the design has been optimized within the constraints imposed on the designer. The components and subsystems are synthesized into the final system in a similar iterative manner.

Evaluation: This step is concerned with measuring the design against the specifications in the problem definition phase. This evolution often requires the fabrication and testing of a prototype model to assess operating performance, quality, and reliability.

Presentation: This step includes documentation of the design by means of drawings, metrices specifications, assembly list, and so on. Essentially, the documentation requires that design data can be created.

7.2 INTEGRATED PRODUCT DEVELOPMENT

A system or product is an assembly of components and subsystems designed to achieve desired functions with acceptable performance and reliability. The characteristics of components, addressed with their reliabilities (quality and life) and associated costs, have a direct effect on the system's reliability. Once the reliability of a system has been determined, engineers are often faced with the task of identifying reliable components in the system in order to optimize the design.

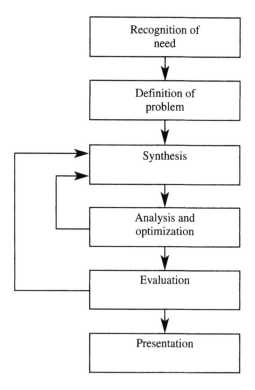

FIGURE 7.1 Conventional design process.

Generally, the designer works within the context of an existing production system that can only be minimally modified. In some cases, the production system will be designed or redesigned in conjunction with the design of the product. When design engineers and production engineers work together to design and rationalize both the product and production and support processes, this is known as integrated product and process design. Whatever the design, analysis of the performance of the product is performed. The starting point for integrated product development is the designer's consideration of:

- Design for manufacturability
- Design for assembly
- Design for serviceability/maintainability
- Design for reliability

7.2.1 DESIGN FOR MANUFACTURABILITY (DFM)

Sometimes products have been designed that could not be produced. An effective product development must go beyond the traditional steps of acquiring and implementing product and process design technology as the solution. It must address the management practices to consider the customer needs, designing those requirements into the product and then ensuring that both the factory and virtual factory (company's suppliers) have the capability to effectively produce the product. Products are initially conceptualized to provide a particular capability and meet identified performance objectives and specifications. Given these specifications, a product can be designed in many different ways. The designer's objective must be to optimize the product design with the production system. This is the basis for design for manufacturability (DFM).

Design for manufacturability, besides considering the customer's needs and designing those requirements into the product, ensures that the factory has the capability to effectively produce the product. The primary objective of the designer is to design a functioning product within the given economic and schedule constraints. Design for manufacturability takes into consideration only the manufacturing aspects of the product and not any other factors. The manufacturing considerations address the assembly efficiency and component complexity. The key in this perspective is to provide the customers with their desired customized products at an affordable price in a timely manner. The design attributes that characterize manufacturing complexity is a commonality of components and process. However, the application of DFM must consider the overall design economics; it must balance the effort and cost associated with the development and refinement of the design to cost and quality leverage that can be achieved. The incorporation of DFM improves production. Designers need to choose an easier way to accomplish the part function. DFM tools and principles provide a structured approach to seeking simplified designs. It focuses mainly on reducing the product complexity. Thus, by simplifying and standardizing designs, establishing design retrieval mechanisms and embedding preferred manufacturing processes in the preferred component list, the design and production efficiencies are enhanced.

7.2.2 DESIGN FOR ASSEMBLY (DFA)

The research in design for assembly (DFA) is based on the premise that the lowest assembly cost can be achieved by designing a product in such a way that it can be economically assembled by the most appropriate assembly system. Three basic types of assembly systems exist, namely, manual, special purpose machine, and programmable machine assembly. The concept of DFA mainly indicates the ratings of each part in the assembly based on the part's ease of handling and insertion. The techniques are concerned with minimizing the cost of assembly within the constraints imposed by design features of the product. The DFA method is summarized as follows:

- Through the use of basic criteria, the existence of each separate part is questioned and the designer is required to provide the reasons why the part cannot be eliminated or combined with others.
- The actual assembly time is estimated using a database of real-time standards developed specifically for the purpose.
- A DFA index (design efficiency) is obtained by comparing the actual assembly time.

- (Comparison of time obtained from standards with actual time taken.)
- Assembly difficulties that may lead to manufacturing and quality problems are identified.
- In the assembly, two factors that influence the assembly cost of the component or subassembly are:
    1. The total number of parts.
    2. The ease of handling, insertion, and fastening of the parts.

Each component's usefulness and functional value are assessed to evaluate the combined rating. This means that the parts that have little functional value, such as separate fasteners, and which are difficult to assemble, are given the lowest ratings. Finally, the ratings are used as guidelines to redesign the products. A spreadsheet approach to rating design on the basis of their ease for automatic assembly is developed. The result shows those component features that tend to increase the assembly costs. The following are the list of DFA criteria:

- Minimize the number of parts and fixings, design variants, assembly movements, and assembly directions.
- Provide suitable lead-in chamfers, automatic alignment, easy access for locating surfaces, symmetrical parts or exaggerated asymmetry, and simple handling and transportation.
- Avoid visual obstructions, simultaneous fitting operations, parts that will tangle or nest, adjustments that affect prior adjustments, and the possibility of assembly errors.

7.2.3 Design for Reliability (DFR)

Reusability is defined as the probability of no failure of a product throughout its operating period. Reliability is the ability of the product to perform its function without any failure during its life cycle period. The reliability of a system, equipment or product is a very important aspect of quality for its consistent performance over its expected lifespan. In the development of new products, the reliability consideration has a tendency to be more of an afterthought in the design process. Reliability activities are performed primarily to satisfy the internal procedures and customer requirements. Mostly, it is done late in the development process. Therefore, efforts must be focused on developing reliability predictions. This effect could be better utilized in understanding and mitigating failure modes, thereby developing improved product reliability. The organizations will undergo repeated and planned design/build/test iterations to develop higher-reliability products. The design of reliability must be involved in the product design at an early point to identify the reliability concerns and begin assessing reliability implications as the design concept emerges.

The design for reliability (DFR) must be done based on the expected range of the operating environment. The reliability is closely related to the life cycle cost of the product. As the reliability of each and every component increases, the manufacturing cost also increases, which leads to a decrease in both service cost and downtime cost.

7.2.4 DESIGN FOR SERVICEABILITY (DFS)

The next perspective addresses the design for serviceability and repair (DFS). The service is done for a product either periodically or at times of failure. For products such as airplanes, locomotives, power generating plants, and manufacturing equipment, the life-cycle operational cost exceeds that of the initial acquisition cost. Some computer peripherals, such as inkjet printers, also exhibit high service in terms of routine replacement of consumable products such as paper and ink cartridges.

The components are designed in such a way that the users must be able to service them easily. The design for serviceability focuses mainly on the functional importance of the product. The key design attribute that characterizes service complexity is the "degree of service difficulty for each service mode." Here service modes refers to:

Routine service/replacement (OTI change, ink cartridge replacement, etc.)
Monitored or scheduled maintenance (bearing replacement, engine overhaul, etc.)
Unscheduled repair (malfunction and accidents)

The key is to provide a design that allows quick and easy high-frequency service operation while allowing reasonable serviceability to low frequency items. Thus, the design for serviceability needs to be developed along with the design of each and every component.

7.3 TOTAL PRODUCT OPTIMIZATION — DESIGN FOR LIFE CYCLE COST (DLCC)

The design optimization process, in general, considers the cost to the customer as the objective criterion. Cost optimization has been attempted with various concepts such as design for manufacturing, design for assembly, design for reliability, and design for serviceability. The above integrated product and process design strategies have been proven more cost effective than the conventional design process that concentrates purely on the functional aspect. However, these considerations aim to optimize any one of the followings costs: production, assembly, reliability, or service.

- Design for production concentrates mainly on component-level man-ufacturing, such as material, process, standardization and simplifica-tion. Components of the system or product are generally designed without considering the product life or reliability. This leads to the production of the components with either higher or lower life and reliability. Both these situations lead to increased cost either in terms of manufacturing or in terms of repair.

- Design for assembly optimizes the assembly cost. On many occasions, an integral part is realized and this leads to replacing the entire part on failure, in case of lower reliability.
- Design for reliability, considered at the system or product level, counts more on redundant or highly reliable components. It may not be a solution for components that could be repaired or replaced with negligible downtime or cost.
- Design for serviceability concentrates on providing service at the lowest cost. It can be considered as the combination of DFA and DFR and leads to unwanted manufacturing costs.

The nature of optimization in each case is very specific to the factors addressed. There is no holistic approach. The effective product cost management requires designing to a holistic cost philosophy.

The cost philosophy should take into account all the cost associated with the performance of the product during its entire life period, called life cycle cost (LCC). The design for minimum LCC is referred to here as total product optimization (TPO). This must involve design for manufacturability, design for assembly, design for serviceability/repair, and design for reliability. The concept of TPO is shown in Figure 7.2.

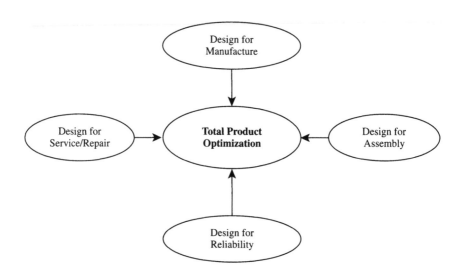

FIGURE 7.2 Total product optimization (TPO).

A review reveals that LCC analysis is an important point to be considered in designing a product and no generalized data model exists combining different factors such as manufacturability, serviceability, and replacement. This chapter presents:

- Modeling for LCC analysis
- Case illustration for an insight to the problem
- A genetic algorithm-based methodology proposed for optimal design with minimum LCC
- Illustration of the proposed methodology
- Future research directions

7.3.1 MODELING FOR LCC ANALYSIS

Various costs involved in owning and using a product for a specified life period are

- Cost of acquisition (manufacturing, assembly, and distribution)
- Cost of servicing and repairing of the product
- Cost of replacing spares and components
- Cost of downtime

The sum of all costs during the lifetime of a product is referred to here as LCC. This chapter considers the design of a product involving all of the above costs. The discussion that follows will reveal that the reliability of its components greatly influences the above costs and, hence, the LCC.

Acquisition cost (AC) comprises the manufacturing cost of all components, assembly, other distribution costs, and profit of the vendor, which is described as:

$$AC = \sum_{i=1}^{m} MC_i + (A + D + P) \tag{7.1}$$

Costs of assembly, distribution, and profit can be assumed as a certain percentage of manufacturing cost, so the acquisition cost can be rewritten as:

$$AC = \left(1 + \frac{C_1}{100}\right) \sum_{i=1}^{m} MC_i \tag{7.2}$$

However, all components are not critical with respect to the reliability of the product. For example, the casing of the CPU of a computer is not critical in the sense that the cost will not be the predominant variation with reliability. On the other hand, the CPU could be designed for various degrees of reliability at different manufacturing costs. In this concern, the components are grouped into two categories as reliability dependent (referred to as critical) and reliability independent (referred to as noncritical), and their characteristics assumed here are:

Critical components:
1. Manufacturing cost varies with reliability and could be designed for different reliability.
2. Requires service/repair during its life cycle.
3. Replaced when its designed life is over.

Noncritical components:
1. No significant variation in cost with reliability and thus designed for the reliability that is equal to the product reliability.
2. Requires no service/repair during its life cycle.
3. No replacement until the end of the product life.

When the product consist of m_1 noncritical components and m_2 critical components, the AC becomes:

$$AC = \left(1 + \frac{C_1}{100}\right)\left(\sum_{j=1}^{m} MC_j + \sum_{k=1}^{m} MC_k\right) \tag{7.3}$$

$$AC = \text{Constant} + \left(1 + \frac{C_1}{100}\right)\left(\sum_{k=1}^{m_2} MC_k\right) \tag{7.4}$$

The reliability R is defined as the probability of survival (P_k) for the intended life (n_k) and hence it is normally denoted with these two parameters. The relationship between acquisition (manufacturing) cost and reliability of a typical critical component is shown in Figure 7.3.

7.3.1.1 Service Cost (SC)

Servicing is inevitable to maintain the operational performance of the product. The product with less reliable components requires very frequent service and, in turn, more service cost (SC). The relationship between service cost and reliability of a critical component is illustrated in Figure 7.4.

The total service cost during the product life period N thus becomes

$$SC = \sum_{k=1}^{m_2} SC_k \tag{7.5}$$

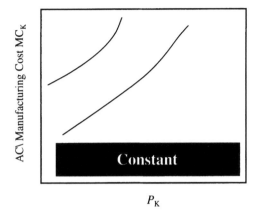

FIGURE 7.3 Relationship between AC (MC_K) and reliability of typical critical component.

where SC_k is the estimated cost of servicing critical component k during the product life, N.

7.3.1.2 Replacement Cost (RC)

Failure of critical components requires replacement. The cost of replacement (RC_k) of critical component k depends on the number of failures F_k during its product life N and the unit cost of replacement (usually proportional to manufacturing cost).

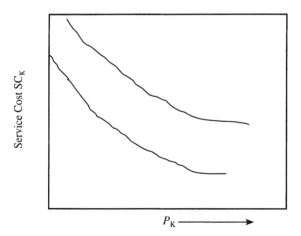

FIGURE 7.4 Variation of service cost with reliability of typical critical component.

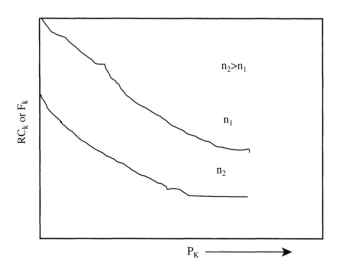

FIGURE 7.5 Variation of cost of failure with reliability of typical critical components.

F_k depends on the component reliability. Unit cost of replacement is assumed as $C_k \times MC_k$. Replacement cost is also dependent on product reliability. The relationship between the cost of failure versus reliability is shown in Figure 7.5.

$$RC_k = \sum_{k=1}^{m_2} F_k \times (C_k MC_k)$$ (7.6)

$$C_k = \text{Constant} > 1$$

$$RC = \sum_{k=1}^{m_2} RC_k$$ (7.7)

7.3.1.3 Downtime Cost (DC)

Downtime (DT) of the product is realized on two occasions:

- The time involved during service of a product.
- The time involved during replacement of failed components. This is usually a function of product reliability R that depends on the component's functionality with the product arrangement (series or parallel).

It is assumed that all critical components affect the functioning of the system and hence the product is considered as a series of components, as shown in Figure 7.6.
The product reliability and thus the multiplication of component reliability is given as:

$$R = \prod_{k=1}^{m_2} R_k$$ (7.8)

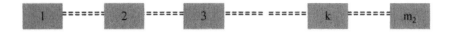

FIGURE 7.6 Series system.

Downtime is a function of R, i.e., $DT = f(R)$. Downtime cost is the multiple of downtime and downtime cost per unit time,

$$DC = C_2 \times DT \tag{7.9}$$

The relationship between downtime cost and reliability of the product is shown in Figure 7.7.

Hence the LCC of the product becomes

$$LCC = \left(1 + \tfrac{C_1}{100}\right)\left(\sum_{j=1}^{m_1} MC_j + \sum_{k=1}^{m_2} MC_k\right) + \sum_{k=1}^{m_2} SC_k$$

$$+ \sum_{k=1}^{m_2} F_k C_k MC_k + DT \times C_2 + f(R) \tag{7.10}$$

$$LCC = \underbrace{\left(1 + \tfrac{C_1}{100}\right)\sum_{j=1}^{m_1} MC_j + \left(1 + \tfrac{C_1}{100}\right)\sum_{k=1}^{m_2} MC_k + \sum_{k=1}^{m_2} SC_k}_{\text{Fixed LCC}}$$

$$\underbrace{+ \sum_{k=1}^{m_2} F_k C_k MC_k + DT \times C_2 + f(R)}_{\text{Variable LCC}} \tag{7.11}$$

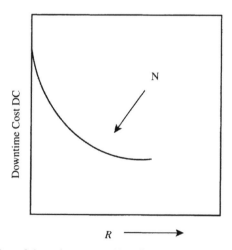

FIGURE 7.7 Variation of downtime cost with reliability of the system.

Equation 7.11 comprises two types of costs: fixed and variable. fixed costs are one-time costs associated with noncritical components and variable costs are associated with critical components that are reliability (P_k, n_k) dependent.

The objective is the minimization of the LCC of the product. As fixed costs do not vary, minimum variable costs lead to a minimum LCC, and hence the objective function is written as:

$$\text{Minimize (LCC}_v) = \left[\sum_{k=1}^{m2} \left(1 + \frac{C1}{100} + Fk * C_k \right) MC_i + SC_k \right] + DT * C2 * f(R)$$

(7.12)

The product is designed for a specific reliability R with the service life of N years and hence the constraint to meet this objective is

$$\prod_{k=1}^{m_2} R_k > R$$ (7.13)

(Noncritical components are assumed not to affect product reliability.)

7.4 CASE ILLUSTRATION

Consider a personal computer that consists of three basic critical subsystems: the motherboard, the hard drive, and a processor. Failure of any one of them would lead to nonavailability of the system and thus it is assumed as a series system. In other words, all the units in this system must succeed for the proper function of this computer.

The computer is to be designed for a lifetime of 5 years with a reliability of 0.80 under 8 hrs/day service requirement. The downtime cost is estimated as Rs. 10/hr. The cost data relevant to the critical components are:

- Manufacturing cost of all critical components for different reliability (P_k, n_k)
- Service cost of all critical components for different reliability (P_k, n_k)

These cost data are known for all the three critical components and their values are given in Table 7.1.

The constant C_1 is taken as 30% and the constants C_k are given as:

k	1	2	3
C_k	1.20	1.27	1.30

TABLE 7.1
Cost Data Relevant to Typical Critical Components

Probability	Cost	Component 1 Life n_1 in Years					Component 2 Life n_2 in Years					Component 3 Life n_3 in Years				
		1	2	3	4	5	1	2	3	4	5	1	2	3	4	5
0.90	MC_k	4050	4100	4200	4500	5000	8050	10,000	10,500	11,000	12,000	4000	4500	5100	5500	6000
	SC_k	1000	870	800	750	700	1000	870	800	750	700	800	760	710	650	600
0.91	MC_k	4700	5250	5750	6100	6500	10,100	11,110	12,200	12,500	13,000	4500	4750	5100	5700	6300
	SC_k	900	850	725	650	600	800	850	725	650	600	750	710	680	640	550
0.92	MC_k	5000	5200	5800	6100	6200	11,000	11,500	11,700	12,500	13,500	4800	5150	5650	5890	6800
	SC_k	850	775	650	550	500	850	775	650	550	500	700	650	610	540	520
0.93	MC_k	5200	5450	5700	6250	6550	11,500	11,900	12,580	13,400	14,200	5200	5600	6100	6800	7100
	SC_k	800	750	675	650	S10	800	750	710	680	680	650	620	590	510	480
0.94	MC_k	5300	5350	5600	5800	6250	12,000	12,560	13,560	14,090	14,500	5650	6150	6800	7150	7550
	SC_k	770	740	625	610	490	750	580	620	590	540	620	590	470	450	420
0.95	MC_k	5400	5500	5700	5850	6300	12,300	12,670	13,700	14,500	14,700	6050	6260	6750	7060	7650
	SC_k	650	610	575	520	440	650	600	540	510	460	610	570	430	420	400
0.96	MC_k	5600	5800	6100	6400	6800	12,700	13,100	13,800	14,500	15,250	6120	6540	6900	7150	7820
	SC_k	600	580	540	500	410	600	540	500	450	400	590	470	450	420	380

(continued)

TABLE 7.1 (Continued)
Cost Data Relevant to Typical Critical Components

Probability	Cost	Component 1					Component 2					Component 3				
		Life n_1 in Years					Life n_2 in Years					Life n_3 in Years				
		1	2	3	4	5	1	2	3	4	5	1	2	3	4	5
0.97	MC_k	5700	5800	6100	6500	6800	13,100	13,600	14,700	15,300	15,680	6200	6580	6980	7250	7900
	SC_k	550	520	400	390	380	550	500	450	400	360	490	460	430	410	390
0.98	MC_k	5900	6100	6300	6500	6900	13,600	14,650	14,800	15,100	15,900	6500	6780	7690	7900	8100
	SC_k	510	490	480	450	360	500	460	410	380	320	480	440	420	395	360
0.99	MC_k	6100	6400	6600	6800	7100	14,050	14,800	15,600	15,700	16,100	6700	6980	7650	7900	8250
	SC_k	480	450	410	380	320	450	410	370	320	310	430	410	380	340	320

TABLE 7.2
Variable Life Cycle Cost

Combination Set	Component 1 P_1	n_1	Component 2 P_2	n_2	Component 3 P_3	n_3	Reliability	Variable LCCv
1	0.91	1	0.92	3	0.97	4	0.812	199,380.55
2	0.98	4	0.94	2	0.91	5	0.838	114,345.60
3	0.96	15	0.96	4	0.95	3	0.875	104,947.30
4	0.92	8	0.91	2	0.98	4	0.620	117,142.36
5	0.93	4	0.93	4	0.92	2	0.796	95,448.00
6	0.93	2	0.92	1	0.96	1	0.821	330,310.60
7	0.94	1	0.98	2	0.93	4	0.857	245,579.25
8	0.95	4	0.99	3	0.92	3	0.865	106,358.20
9	0.93	3	0.95	4	0.93	2	0.822	138,423.40
10	0.96	2	0.98	5	0.94	3	0.884	128,886.60

With the above data set, the problem is to determine the reliability of each component that would lead to a minimum life cycle cost. The variable LCC for a few component set (different reliability combinations) is calculated and given in Table 7.2.

Among the 10 sets shown above, Set 5 provides the minimum variable life cycle cost. But this does not satisfy the product reliability requirement (i.e., reliability of 0.80) and hence it is not an acceptable solution. The next minimum cost, corresponding to Set 3, satisfies the product reliability condition and thus is acceptable. However, the above 10 sets are only a small percentage of the total solution space (combination). An optimum solution could be obtained by enumerating all possible combinations.

7.5 PROPOSED METHODOLOGY

The illustration given in the previous section shows that this combinatorial optimization problem requires efficient heuristic algorithms as an alternate to complete enumeration. A genetic algorithm is outlined below.

Step 1: Generation of initial population. A set of chromosomes equal to *pop_size* (assumed as 10) is generated initially by random choice. Each chromosome (c) is the representation of possible combination of reliabilities (P_k, n_k) of critical components (m_2). Hence, the number of genes in a chromosome is equal to $2m_2$. The odd genes represent P_k and the even genes represent n_k in such a manner that the kth odd/even gene indicates the values for the kth critical components in their ranges. The *pop_size* is problem solution-space dependent; *pop_size* is usually higher for a higher solution space.

Step 2: Evaluation of chromosomes. The chromosomes are evaluated with LCC_v (Equation 7.12) that becomes the fitness parameter, i.e., fit(c) = LCC_v.

Step 3: Identification of best chromosome, based on the fitness parameter. Identify the best chromosome of the current population and check whether it is better than the previously stored best. If so, store this value as the best chromosome; otherwise, leave the previous best as best.

Step 4: Check for termination. This step terminates the program by checking the generation number or solution convergence. In this problem, the maximum number of generations is set as the termination criteria (assumed as 100). When the generation number becomes the maximum number of generations, the program terminates and prints the best chromosome as the output. Otherwise, proceed to the next step.

Step 5: New population generation. This involves three steps: Selection, Crossover, and Mutation.

Step 5.1: Selection module. A new population of same *pop_size* is selected based on the probability of survival of chromosomes with a modified fitness parameter given as

$$\text{Mod-fit}(c) = e^{-x \cdot \text{fit}(c)} + \lambda$$

$$\lambda = 0 \text{ for chromosome } \prod_{k=1}^{m_2} R_k \geq \prod R$$

The negative exponential function takes care of the minimization objective. The scaling constant x is suitably selected such that good chromosomes have more repetitions (multiple copies) and worse chromosomes have fewer copies in the new set.

Step 5.2: Crossover module. Many crossover operators are addressed in the literature. As the problem involves a limited number of critical components and crossover must be done with reliability sets (i.e., P_k, n_k), a single-point crossover operation with an edge crossover operator is employed in this problem, as explained below.

Potential chromosomes for crossover are selected with a probability of crossover *p_cross* (assumed here as 0.6). Crossover with a set of chromosomes is carried out by generating a random number r between 1 and m_2 and the genes after $2r$ are exchanged and maintain feasibility.

Step 5.3: Mutation module. A random number r_1 between 0 and 1 is generated for each gene of the entire population, and the gene that gets a random number less than the probability of mutation *p_mut* (assumed here as 0.05) is mutated with some other value of P_k or n_k, as the case may be.

Step 6: Go to Step 2.

7.6 GA ILLUSTRATED

The proposed methodology is illustrated with the example problem given earlier in this chapter.

Step 1: The initial population generated of size 10 is given in columns 1 and 4 of Table 7.2. The first column represents chromosome number, c. The second column represents (P_1, n_1) values of component 1. The third column represents (P_2, n_2) values of component 2. The fourth column represents (P_3, n_3) values of component 3.

Step 2: The evaluation parameter of LCC_v for all chromosomes is calculated and is given in column 6 of Table 7.2.

Step 3: The chromosome 3 that satisfies the reliability requirement condition and has the least LCC_v becomes the best chromosome in this generation and is stored as the best combination.

Best chromosome 3:

Chromosome	Component 1		Component 2		Component 3		Reliability	Variable LCC
c	P_1	n_1	P_2	n_2	P_3	n_3	R	LCC_v
3	0.96	3	0.96	4	0.95	3	0.875	104,947.30

Step 4: Because this is the first generation, the process continues until the best minimum occurs.

Step 5: The chromosome selected with the *mod_fit* function is given as Step 4. Because this is the first generation, the process continues until the best minimum occurs.

Step 6: The chromosome selected with the *mod_fit* function is given as New Population c.

Selected chromosome (c')	1′	2′	3′	4′	5′	6′	7′	8′	9′	10′
Old chromosome (c)	3	8	3	2	4	10	3	6	2	9

The chromosomes selected for crossover with a *p_cross* value of 0.5 are 12 (3), 42 (2), 72 (3), and 82 (6). The offspring of the above chromosome are 1, 4, 7, and 8. The following table shows the chromosomes before crossover and after crossover with random r.

Old Chromosome		Before Crossover						R	After Crossover						New Chromosome
		G1	G2	G3	G4	G5	G6		G1	G2	G3	G4	G5	G6	
Set 1	12	0.96	3	0.96	4	0.95	3		0.96	3	0.96	4	0.91	5	13
								2							
	42	0.98	4	0.94	2	0.91	5		098	4	0.94	2	0.95	3	43
Set 2	72	0.96	3	0.96	4	0.95	3	2	0.96	3	0.96	4	0.96	1	73
	82	0.93	2	0.92	1	0.96	1		0.93	2	0.92	1	0.95	3	83

The new population after crossover thus becomes 1″, 2′, 3′, 4″, 5′, 6′, 7″, 8″, 9′, and 10′. The new population after mutation (full generation) with p_mut = 0.5 is given in the following table:

Chromosome c	Component 1		Component 2		Component 3		Reliability	Variable LCC
	P_1	n_1	P_2	n_2	P_3	n_3	R	LCC_v
1 (1″)	0.96	3	0.96	4	0.91	5	0.839	88,057.30
2 (2′)	0.98	4	0.94	2	0.91	5	0.838	114,345.60
3 (3′)	0.96	3	0.96	4	0.95	3	0.875	104,947.30
4 (4″)	0.98	4	0.94	2	0.95	3	0.875	104,595.80
5 (5′)	0.93	4	0.93	4	0.92	2	0.796	95,448.00
6 (6′)	0.93	2	0.92	1	0.96	1	0.821	330,310.60
7 (7″)	0.96	3	0.96	4	0.96	1	0.885	219,772.00
8 (8″)	0.93	2	0.92	1	0.95	3	0.815	206,752.00
9 (9′)	0.93	3	0.95	4	0.93	2	0.822	138,423.40
10 (10′)	0.96	2	0.98	5	0.94	3	0.884	128,886.60

This process of evaluation and new population generation is carried out 100 times and the optimal solution obtained at the 67 th Generation is given below:

Component 1		Component 2		Component 3		Reliability	Variable LCC
P_1	n_1	P_2	n_2	P_3	n_3	R	LCC_v
0.96	3	0.96	4	0.91	5	0.839	88,057.30

7.7 CONCLUSION

In this chapter, the concept of total product optimization for product design considering LCC that associates performance specifications with least life cycle cost is presented. The concept and methodology are well illustrated with a simple case analysis. However, its application requires correct individual critical component reliability values that meet the system reliability while satisfying the associated performance specifications and while the least life cycle costs are presented. The future work must concentrate on the development of empirical curves to predict those values with CAD modeling and dynamic analysis. Upon the evolution of this, a new concept of total product optimization will feature prominently in CAD packages.

NOMENCLATURE

A	Assembly cost of product
C	Chromosome
C_1	Cost of assembly and distribution and profit expressed as percentage of manufacturing cost
C_2	Cost of downtime per hour
C_k	Proportionality constant of component for replacement cost
D	Distribution cost of product
DC	Total downtime cost
DT	Downtime of the component during entire life N
F_k	Number of failures of component k
i	Component identifier
j	Noncritical component identifier
k	Critical component identifier
LCC	Life cycle cost
LCC_v	Variable life cycle cost
m	Number of components
m_1	Number of noncritical components
m_2	Number of critical components
MC_j	Manufacturing cost of component i
MC_k	Manufacturing cost of critical component k
N	Life of the product
n_k	Life of component k
P	Profit of product
p_cross	Probability of crossover
p_mut	Probability of mutation
P_k	Probability of survival of component k
pop_size	Population size
r, r_1	Random numbers
RC_k	Replacement cost of component k during entire life N
SC	Total service cost of the product during entire life N
SC_k	Service cost of component k during entire life N
x	Constant

REFERENCES

Hsu, W., and Lee, C. S. G., Fuzzy Application in Tolerance Design, *Proc. of IEEE International Conference on Fuzzy Systems,* Orlando, Florida, June 26–July 2, 1994.

Nachtmann, H., and Chimka, R.J., Fuzzy reliability in conceptual design, *Proceedings of the Annual Reliability and Maintainability Symposium,* University of Arlcansas, Fayelteville, AR, Vol. 29, 360–364, 2003.

Patterson, J.L. and Dietrich, D., Dirichlet binomial attribute testing model: A Bayesian approach to estimating reliability decay, *Proceedings of the Annual Reliability and Maintainability Symposium,* Ottawa, Canada, 393–400, 2001.

Tsai, C.K., Hong, S.H., and Zhang, H.C., *Design for Manufacturability and Design for X: Concepts, Applications and Perspectives,* Elsevier Science Ltd., Burlington, VT, 241–260, 2001.

Vintr, Z., Optimization of reliability requirements from manufacturer's point of view, *Proceedings of the Annual Reliability and Maintainability Symposium,* Miltary Academy, Brno, Czech Republic, 183–189, 2001.

8 Scheduling Optimization

Scheduling can be described as the allocation of available resources over time to meet the performance criteria defined in a domain. Typically, scheduling handles a set of jobs to be completed and each job consists of a set of operations. Each operation is performed by specific resources, such as machines and operators. In terms of scheduling theory, most scheduling problems are in the class of NP-hard. Scheduling aims at a proper allocation of resources to tasks to minimize cost and maximize profit. Resources are workers, machines, and material handling; tasks are jobs and services; output is the time that each job starts and completes in each work station and machine. Scheduling plays a crucial role not only in the efficiency of operating the system, but also in customer satisfaction.

Due to increasing market competition, companies strive to:

- Shorten delivery times
- Increase variety in end-products
- Shorten production lead times
- Increase resource utilization
- Improve quality and reduce WIP
- Prevent production disturbances (machine breakdowns)
- Produce more products in less time

The effective scheduling of the tasks to the resources can achieve all of this.

8.1 CLASSIFICATION OF SCHEDULING PROBLEMS

A specific scheduling problem is described by four types of information:

- The jobs and operations to be processed
- The number and types of machines that comprise the shop
- Disciplines that restrict the manner of making assignments
- The criteria to evaluate a schedule

Static and dynamic scheduling problems are classified by the nature of job arrivals. In static scheduling problems, a certain number of jobs arrive simultaneously in a shop that is idle and immediately available for work. No further jobs will arrive, so attention can focus on scheduling this completely known and available set of jobs. In dynamic scheduling problems, the shop is a continuous process. Jobs arrive intermittently at times that are only in a statistical sense and arrivals will continue indefinitely into the future.

8.1.1 SINGLE MACHINE SCHEDULING

The simplest pure sequencing problem is one in which there is a single resource, or machine. As simple as it is, however, the single machine case is still very important for several reasons. It provides a context in which to investigate many different performance measures and several solution techniques. It is therefore a building block in the development of the comprehensive understanding of the scheduling concepts, an understanding that ultimately facilitates the modeling of complicated systems.

8.1.2 FLOW SHOP SCHEDULING

Any group of machines served by a unidirectional, noncyclic conveyor is considered a flow shop. A flow shop is one in which all the jobs follow essentially the same path from one machine to another. In the flow shop, machines are arranged in series and jobs begin processing on an initial machine, proceed through several intermediary machines, and are completed in the final machine. Each machine will take up the jobs in a sequence to perform the operation required. The sequence of jobs for all the machines is the same.

8.1.3 SCHEDULING OF JOB SHOPS

In a job shop, a specific machine order restriction is not imposed on each job. In this scenario, a job can be processed on machines in any order. The general job shop is one in which n jobs are to be processed by m machines. Each job will have a set of constraints on the order in which machines can process and a given processing time on each machine. Jobs may not require all m machines and they may have to visit some machines more than once.

Only one machine can perform a given operation. In practice, often multiple copies of the same machine exist (parallel machines), so a number of machines can process the jobs. This adds more flexibility in a shop but it complicates the scheduling problem even further. In the job shop model, all the jobs need not have the same technological order of operations. Each job can have different routes. In the job shop-type problem, not only the estimated processing times of jobs in the various machines are required, but also the route the job has to take.

8.1.4 PARALLEL MACHINE SCHEDULING

Parallel machine scheduling (PMS) is used to schedule jobs processed on a series of same function machines in order to achieve certain objective functions. If N jobs must be processed on M machines and if the processing requirement p_{ij} satisfies $p_{ij} = p_j$ for any M_i, where jobs ($j = 1, 2, 3...N$) and machines ($i = 1, 2, 3...M$), then they are identical parallel machines.

The scheduling process involves two kinds of decisions, sequencing and job-machine assignment. The complexity usually grows exponentially with the number of machines, making the problem intractable. This problem, like all deterministic scheduling problems, belongs to the wide class of combinatorial optimization problems, many of which are known to be NP-hard. PMS is a manufacturing philosophy that leads to achieving the just-in-time (JIT) production system by eliminating bottleneck operations. The adaptation of a PMS in a manufacturing environment has the advantages of simplified material flow, reduced work-in-process inventory, and better control of manufacturing activities.

8.1.5 FMS Scheduling

In a flexible manufacturing system (FMS) life cycle, different levels of decision problems exist for its design and operation. Among them, scheduling is a critical function for the control and operation of any FMS. Scheduling in flexible manufacturing systems differs from that of a conventional job shop because any one of several machines can perform each operation of a job. FMS scheduling has been researched extensively over the past two decades and most FMS scheduling problems have been shown NP-hard. Scheduling in FMS is more complex than scheduling in classical machine shops. The additional complexity comes from the following two factors, together with the need for real time operation.

> In FMS scheduling, decisions that need to be made include not only sequencing of jobs on machines, but also the routing of the jobs through the system. Apart from the machines, other resources in the system, e.g., material-handling devices, must be considered.

Scheduling in flexible manufacturing systems should include the following three aspects to exploit the potential of system flexibility:

- Machine setup or tool changing
- Part routing
- Operation sequencing

In solving the FMS scheduling problem, decisions on these two aspects must be made to achieve some scheduling objectives, such as maximizing machine utilization and minimizing the penalty cost. However, before the sequencing decision is made, precisely measuring the performance of these two decisions is impossible. The sequencing decision, on the other hand, cannot be made without the specification of part routing at the same time or in advance. Therefore, all three aspects are closely related. These interconnections and the additional resource constraints make the scheduling problem in FMS more complex than those in conventional machine shops. Ideally, considering all three aspects of the scheduling decisions and the constraints of all the resources in the system concurrently should solve the FMS scheduling problem.

8.2 SCHEDULING ALGORITHMS

General solution techniques:
Mathematical programming

- Linear, nonlinear, (mixed) integer programming
- Branch-and-bound
- Dynamic programming
- Cutting plane and column generation methods

Heuristics

- Dispatching rules
- Composite dispatching rules
- Beam-search

Intelligent techniques

All the intelligent techniques described in Chapter 3 can be applied to solve the different scheduling problems. In this section, application of genetic algorithm and simulated annealing algorithm are illustrated for solving parallel machine and flow shop scheduling, respectively.

8.3 PARALLEL MACHINE SCHEDULING OPTIMIZATION USING GENETIC ALGORITHM

8.3.1 Data of Problem

Number of machines: 3 identical machines
Number of jobs: 10
Working hours: 8 hours/day

Job No.	Processing Time (min)	Due Date (Days)	Batch Quantity
0	2	11	218
1	1	07	112
2	7	12	711
3	6	13	655
4	4	03	419
5	3	12	354
6	1	03	174
7	9	07	910
8	2	08	076
9	1	10	249

8.3.2 Genetic Algorithm Parameters

Population size: 30
Length of chromosomes: 10
Selection operator: Rank order
Crossover operator: Single point operator
Crossover probability (P_c): 0.6
Mutation probability (P_m): 0.01

8.3.3 Fitness Parameter

Minimization of makespan (total time to complete all the jobs) is considered as objective function. Ten available jobs are scheduled on three identical parallel machines. The first available machine is selected for operation.

8.3.4 Representation

Representation plays a key role in the development of a GA. A problem can be solved once it can be represented in the form of a solution string (chromosomes). The bits (genes) in the chromosomes could be binary or real integer numbers. In this example, the job sequences are considered the primary parameters for the scheduling functions. There are 10 jobs and each of these jobs is represented in real integer numbers from 0 to 9.

8.3.5 Initialization

During initialization, a solution space of a "population size" solution string is randomly generated. The solution space size is considered to be 30.

8.3.6 Crossover

Crossover is a mechanism for diversification. The strings to be crossed and the crossing points are selected randomly and crossover is done with a crossover probability. A single point crossover is used in this work. The crossover probability is 0.6. The concept of crossover is explained below.

Before crossover:

8-2-6-1-3-0-9-7-4-5
9-8-0-5-1-6-2-4-3-7

A crossover site is selected at random. For example, for the above pair the generated number is 2. After the second bit, all the information is exchanged between strings (i.e., rearranging the numbers of the first schedule as per the second schedule and rearranging the numbers of second schedule as per the first).

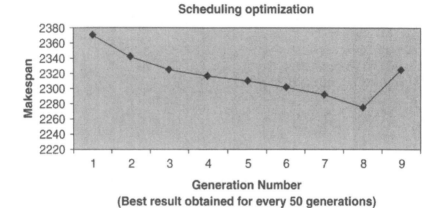

FIGURE 8.1 Scheduling optimization results using GA.

After crossover:

8-2-9-0-5-1-6-4-3-7
9-8-2-6-1-3-0-7-4-5

8.3.7 MUTATION

Mutation is a random modification of a randomly selected string. Mutation is done with a mutation probability of 0.01. Strings 11 and 21 are mutated.

Before Mutation: 1-2-8-0-4-5-3-6-7-9
After Mutation: 1-2-8-0-4-5-3-7-6-9

For the first schedule, generated random numbers are 8 and 9 (i.e., the 8th bit is shifted to the 9th position and the 9th bit is shifted to the 8th position). The solution history is given in Figure 8.1.

Best result from the initial random schedule:
7-1-5-10-9-4-8-6-3-2 (Makespan: 2370 min)
Best result from the schedule generated during 500 generations:
10-8-6-4-5-1-3-9-2-7 (Makespan: 2275 min.)

8.4 IMPLEMENTATION OF SIMULATED ANNEALING ALGORITHM

8.4.1 NOTATIONS AND TERMINOLOGY

FR_CNT: Freeze counter used to check whether the algorithm could be frozen or not. When the counter reaches 5, the algorithm is deemed to be frozen; the counter is reset to zero whenever we find that $\tau_s \leq \tau_B$.

ACCEPT: Counter to keep track of the number of accepted moves at a particular temperature.

TOTAL: Counter to keep track of the total number of moves at a particular temperature.

PER: Stores the percentage of accepted moves at a particular temperature.

B: The best sequence obtained so far. Whenever B is updated by a better solution, FR_CNT is reset to zero with the renewed hope of obtaining a still better solution later.

S: Sequence on which RIP/CRIP (random insertion perturbation scheme/curtailed RIP) is employed.

S: Sequence that is obtained by perturbing S (i.e., by employing RIP on S).

τ_B, τ_S, $\tau_{S'}$: Sum of tardiness of jobs for sequences, B, S, and S', respectively.

RIPS(S, S', $\tau_{S'}$): The routing using RIPS for which sequence S is the input and S' is the output with the sum of tardiness of jobs given by $\tau_{S'}$.

CRIPS(S, S', $\tau_{S'}$): The routine using CRIPS for which sequence S' is the input and S is the output with the sum of tardiness of jobs given by $\tau_{S'}$.

DEL(B, S'): The routine for computing the relative improvement (or deterioration) of S' with respect to B.

DELTA_B $= (\tau_{S'} - \tau_B) \times 100/\tau_B$

DEL(S, S'): The routine for computing the relative improvement (or deterioration) of S with respect to S.

$\Delta = (\tau_{S'} - \tau_S) \times 100/\tau_S$

P: Acceptance probability given by $e^{\Delta/T}$

U: Uniform random number that is used to check for the acceptance of an inferior solution.

8.4.2 SA ALGORITHM WITH RIPS: STEP-BY-STEP PROCEDURE

Step 1: Obtain the initial sequence by using random method. Assign this sequence to both S and B. Compute the sum of tardiness of jobs in sequence S and B assign it to both τ_S and τ_B.

Step 2: Initialize $T = 475$, ACCEPT $= 0$, TOTAL $= 0$, and FR_CNF $= 0$.

Step 3: If (FR_CNT $= 5$) or ($T \leq 20$), go to Step 14; else, proceed to Step 4.

Step 4: Invoke the perturbation routine RIPS(S, S', $\tau_{S'}$), and hence compute Δ from DEL(S, S'). If $\Delta \leq 0$, proceed to Step 5; else, go to Step 8.

Step 5: Assign $S \Leftarrow S'$, $\tau_S \Leftarrow \tau_{S'}$ and ACCEPT $=$ ACCEPT $+ 1$.

Step 6: Invoke the routine DEL(B, S'), and hence compute DELTA_B. If DELTA_B ≤ 0, proceed to Step 7; else, go to Step 10.

Step 7: Assign $B \Leftarrow S'$, $\tau_B \Leftarrow \tau_{S'}$, and FR_CNT $= 0$, and go to Step 10.

Step 8: Compute P and sample U. If $U > P$, go to Step 10; else, proceed to Step 9.

Step 9: Assign $S \Leftarrow S$, $\tau_S \Leftarrow \tau_S$ and ACCEPT $=$ ACCEPT $+ 1$.

Step 10: Set TOTAL $=$ TOTAL $+ 1$.

Step 11: If (TOTAL > 2n) or (ACCEPT > n/2), proceed to Step 12; else go to Step 4.

Step 12: Compute PER = (ACCEPT × 100/TOTAL). If PER ≤ 15, set FR_CNT = FR_CNT + 1, and proceed to Step 13; else, proceed to Step 13.

Step 13: Set $T = T \times 0.9$, ACCEPT = 0, TOTAL = 0, and go to Step 3.

Step 14: The algorithm is frozen. B contains the heuristic sequence and τ_B contains the sum of tardiness of jobs in the sequence; therefore, stop.

8.4.3 SA Algorithm with CRIPS

All steps in the SA algorithm with RIPS remain the same except for Step 4, which is to read as follows:

Step 4: Invoke the perturbation routine CRIPS(S, S', $\tau_{S'}$), and hence compute Δ from DEL(S, S'). If $\Delta \leq 0$, proceed to Step 5; else, go to Step 8.

8.4.4 Numerical Illustration

Consider a flowshop with five jobs and three machines with the following details and the objective of minimizing the sum of job tardiness:

	Machine j			
Job i	1	2	3	Due Date (d)
1	11	94	32	187
2	32	26	97	196
3	42	13	24	180
4	20	20	97	215
5	10	1	36	119

8.4.5 Obtaining Seed Sequence

By using the random method, the following three sequences are obtained: {1-5-4-2-3}, {5-4-2-3-1}, and {5-3-1-2-4} with the sum of job tardiness values as 491, 249, and 266, respectively. Hence, the sequence {5-4-2-3-1} is chosen as the seed sequence.

8.4.6 Improvement by SA Algorithm with RIPS

Assign the seed sequence {5-4-2-3-1} to both S and B with their τ_S and τ_B values as 249 (Step 1). Set $T = 475$, ACCEPT = 0, TOTAL = 0, and FR_CNT = 0 (Step 2). Because the freeze condition is not satisfied, proceed to Step 4.

Perturb the sequence S using the RIPS. Note that the RIPS generates eight sequences and returns the best among them as S, in this case, the sequence {5-4-3-1-2} with $\tau_S = 136$. Compute

$$\Delta = [(136 - 249) \times 100/249]$$

The negative value shows that the perturbed solution is superior. Hence, set $S \Leftarrow S$ and $\tau_S = \tau_{S'}$. Increment ACCEPT by one (Step 5).

Making use of B and S, calculate DELTA_B. The negative value shows that the perturbed solution is better than the best solution obtained so far, hence set $B \Leftarrow S'$ and $\tau_B = \tau_{S'}$ and reset FR_CNT = 0 (Step 7). Increment TOTAL by one (Step 10). Return to Step 4 as the condition in Step 11 is not satisfied, and the processes of perturbation and evaluation of schedules are again carried out. This process continues until the condition in Step 11 is satisfied, after which compute

$$\text{PER} = [(\text{ACCEPT} \times 100)/\text{TOTAL}] \ (\text{Step } 12)$$

If PER $\leq$ 15, then set $T = T \times 0.9$ and reset ACCEPT and TOTAL to zero. Return to Step 3. The procedure is continued until the "freeze condition" is satisfied. In this example, the best sequence obtained is $B\{5\text{-}4\text{-}1\text{-}3\text{-}2\}$ with $\tau_B = 127$.

REFERENCES

Baker, K.R., *Introduction to Sequencing and Scheduling*. John Wiley & Sons, New York, 1974.

Balasubramaniam, H., Monch, L., Fowler, J., and Pfund, M., Heuristic scheduling of jobs on parallel batch machines with incompatible job families and unequal ready times, *International Journal of Production Research*, 42(18), 1621, 2004.

Cheng, R. and Gen, M., Parallel machine scheduling problems using memetic algorithms, *Computers and Industrial Engineering*, 33(3-4), 761–764, 1997.

Cheng, R., Gen, M., and Tozawa, T., Minimax earliness/tardiness scheduling in identical parallel machine system using Genetic Algorithms, *Computers and Industrial Engineering*, 1(4), 513–517, 1995.

French, S., *Sequencing and Scheduling: An Introduction to the Mathematics of the Job-Shop*, Ellis Horwood, Chichester, UK, 1982.

Ishibuchi, H., Misaki, S., and Tanaka, H., Modified simulated annealing algorithms for the flow shop sequencing problems, *European Journal of Operational Research*, 81, 388–398, 1995.

Johnson, D.S., Argon, C.R., McGeoch, L.A., and Schevon, C., Optimisation by simulated annealing: an experimental evaluation, part 1 — graph partitioning, *Operations Research*, 37, 865–891, 1989.

Kim, Y.D., Heuristics for flowshop scheduling problems minimizing mean tardiness, *Journal of Operational Research Society*, 44, 19–29, 1993.

Kirkpatrick, S., Gelatt, C.D., Jr., and Vecchi, M.P., Optimization by simulated annealing, *Science*, 220, 671–680, 1983.

Logendran, R. and Nudtasomboon, N., Minimizing the makespan of a group scheduling problem: a new heuristic, *International Journal of Production Economics*, 22, 217–230, 1991.

Min, L. and Cheng, W., A genetic algorithm for minimizing the makespan in the case of scheduling identical parallel machines, *Artificial Intelligence in Engineering*, 13(4), 399–403, 1999.

Mokotoff, E., Parallel machine scheduling: a survey, *Asia-Pacific Journal of Operational Research*, 18(2), 193, 2001.

Ogbu, F.A. and Smith, D.K., The application of the simulated annealing algorithm to the solution of the flowshop problem, *Computers and Operations Research,* 17, 243–253, 1990.

Rajendran, C., Heuristic algorithm for scheduling in a flowshop to minimize total flow time, *International Journal of Production Economics,* 29, 65–73, 1993.

Sridhar, J. and Rajendran, C., Scheduling in a cellular manufacturing system: a simulated annealing approach, *International Journal of Production Research,* 31, 2927–2945, 1993.

Van Laarhoven, P.J.M. and Aarts, E.H.L., *Simulated Annealing: Theory and Applications,* Reidal, Dordrecht, Netherlands, 1987.

9 Modern Manufacturing Applications

9.1 IMPLEMENTATION OF GENETIC ALGORITHM FOR GROUPING OF PART FAMILIES AND MACHINING CELL

Cellular manufacturing is the application of group technology in which similar parts are identified and grouped together to take advantage of the similarities in the part geometries and operation processes in the manufacturing process. By grouping several parts into part families based on either geometric shape or operation processes, and also forming machine groups or cells that process the designated part families, reducing costs and streamlining the work flow is possible. Cellular manufacturing enables economies of scale approaching those of mass production, enhanced standardization of parts and processes, and elimination of duplication in design proliferation of different part routings on the shop floor.

In this chapter, a 30-machine and 16-component problem is examined and the implementation of GA is described for the concurrent formation of part families and machine groups for cellular manufacturing.

9.1.1 DATA OF PROBLEM

The machine part incident matrix is given in Table 9.1.

9.1.2 CODING SCHEME

A 46 digit number can be used to code the formation of part families and machine groups,

11000100 10101101 01001001 11010000 10011111 111111

By decoding the above number, the following grouping can be obtained by considering two-part families and two machine groups:

Part family one: 1, 2, 6, 9, 11, 13, 14, 16 (8 components)
Machine group one: 2, 5, 8, 9, 10, 12, 17, 20, 21, 22, 23, 24, 25, 26, 27, 28, 29, 30 (18 machines)

The remaining parts and machines form the second part family and second machine group.

The first 16 digits represent the part family and the next 30 digits represent the machine group. The first position (1) means component 1 belongs to the first part family; if 0, it belongs to the second part family. Similarly, if the 17th digit is 1, this means machine 1 belongs to the first machine group, or else it belongs to the second machine group. Likewise, the number can be decoded and the corresponding grouping can be obtained.

9.1.3 CROSSOVER OPERATION

Normally, crossover is performed between a pair of binary numbers. But with this coding system, crossover is required to be performed within the sample,

Group 1: 11000100 10101101 01001001 11010000 10011111 111111
Group 2: 00111011 01010010 10110110 00101111 01100000 000000

To perform the crossover, generate a random number between 1 and 46. If the number is between 1 and 16, cross the binary number representing the parts, or else cross the binary number representing the machines. For example, if the random number is 10, this means to exchange the digits after 10 between the binary number representing the parts for Groups 1 and 2. If it is 20, exchange the digits after 4 between the binary number representing the machines for Groups 1 and 2.

9.1.4 MUTATION

Decide the number of random numbers according to mutation probability and generate numbers in the range of 1 to 46. According to the random number, select a particular bit from the parts or machines and exchange that number between the Groups 1 and 2. For example, if the random number is 7, select the 7th bit from the binary number representing the parts for Groups 1 and 2 and interchange this particular digit. If the random number is 40, select the 24th bit from the binary number representing the machines for Groups 1 and 2 and interchange this particular digit.

9.2 SELECTION OF ROBOT COORDINATE SYSTEMS USING GENETIC ALGORITHM

The motion of an industrial robot manipulator is generally specified in terms of the motion of the end effector in Cartesian space. An accurate direct measurement of the end effector position is a complex task, and the implementation of a motion control system in Cartesian space can be very difficult. Thus, in practical cases the end effector's motion is converted into joint motion using inverse kinematics and then the control task is performed in joint space.

In forward kinematics, if the joint variables are available, the location of the robot can be determined using the equations of motion. In reverse kinematics, the joint variables must be determined to place the robot at the desired location.

TABLE 9.1
Machine-Part Incident Matrix (16M × 30P)

Problem No = 2

	Parts																													
Machines	1	2	3	4	5	6	7	8	9	10	11	12	13	14	15	16	17	18	19	20	21	22	23	24	25	26	27	28	29	30
1						1		1	1	1	1	1	1					1			1		1		1	1	1			
2					1	1	1		1	1	1	1	1		1								1	1		1		1		
3	1	1	1	1											1	1	1													
4			1		1		1	1	1	1								1			1		1	1	1	1	1	1	1	1
5		1			1	1	1	1	1	1										1	1	1	1	1	1	1	1	1	1	1
6		1				1	1													1			1	1	1	1	1	1	1	1
7							1	1	1	1	1	1	1						1	1	1				1	1				
8				1	1			1	1	1	1	1	1							1	1		1	1	1	1	1			
9											1																			
10	1	1		1	1			1	1	1	1	1		1		1			1							1	1			
11					1			1	1	1	1	1	1	1				1	1	1	1		1							
12			1														1	1		1	1	1	1					1	1	1
13			1				1											1		1	1			1				1	1	1
14					1		1	1	1	1	1	1				1														
15	1													1	1															
16	1	1												1		1														

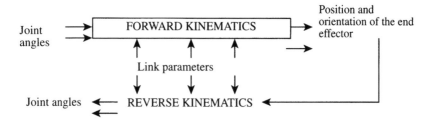

FIGURE 9.1 Forward and reverse kinematics of robot.

9.2.1 THREE-DEGREES-OF-FREEDOM ARM IN TWO DIMENSIONS

Figure 9.1 shows the robot manipulator with the three-jointed links as its abstract model. The workspace is constrained by the combined length of the three links and by a maximum rotational displacement limit of 180° at the elbow joint/base. The origin of the world system is assumed to be at the robot base because the objects would be placed at the horizontal plane passing through the base, making $z = 0$. Because the desired state of the robot is normally specified by the position of the end effector, an accurate and flexible coordinate transformations capability from the Cartesian location to corresponding joint angles is needed.

The position of the end effector on the x-y plane is determined numerically from the following relations:

$$X = [L_2 \cos (\alpha - 90) + L_3 \sin\beta] \cos \theta$$

$$Y = [L_2 \cos (\alpha - 90) + L_3 \sin\beta] \sin \theta$$

L_1, L_2, L_3 = Length of the arms/links
θ, α, β = Angles formed by links L_1, L_2, L_3

9.2.2 THREE-DEGREE-OF-FREEDOM ARM IN THREE DIMENSIONS

Figure 9.2 shows the robot manipulator with the three-jointed links as an abstract model. The workspace is constrained by the combined length of the three links and by a maximum rotational displacement limit of 180° at the elbow joint/base. The origin of the world system is assumed to be at the robot base because the objects would be placed at the horizontal plane passing through the base, making $z = 0$. Because the desired state of the robot is normally specified by the position of the end effector, an accurate and flexible coordinate transformations capability from the Cartesian location to corresponding joint angles is needed.

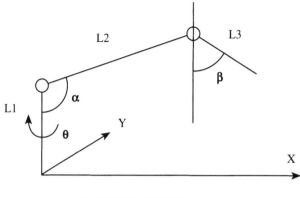

θ, α, β - JOINTS ANGLES
L1,L2,L3 – ARMS LENGTH

FIGURE 9.2 Three-joint robot.

The position of the end effector on the x-y-z plane is determined numerically from the following relations:

$$X = \cos \theta \, (L \cos \Phi + L_4 \cos \Phi)$$

$$Y = \sin \theta \, (L \cos \Phi + L_4 \cos \Phi)$$

$$Z = L_1 + L \sin \Phi + L_4 \sin \Phi$$

L_1, L_2, L_3 = Length of the arms/links
θ, Φ, Φ = Angles formed by links L_1, L_2, L_3

9.2.3 OBJECTIVE FUNCTION

To minimize the error (difference between the required coordinates and the obtained coordinates).

$$\text{Minimum error} = (X - X_{req})^2 + (Y - Y_{req})^2$$

X, Y = Obtained value of x, y
X_{req}, Y_{req} = The value of x, y that is desired

For determination of an (X, Y) point in the space from the robot. Various points were taken as input. The input values of X, Y are the desired locations the end effector has to reach.

The objective is to reduce the error to the least possible value. The genetic algorithm has been used to solve this problem and it serves two purposes: to find the end position as accurately as possible, and to reach the desired point as quickly as possible by quick computation.

9.2.4 INPUT DATA FOR TWO-DIMENSION PROBLEM

Length of links	Range of angles
$L_1 = 60$ cm	$\alpha = 77°$ to $114°$
$L_2 = 50$ cm	$\beta = 0°$ to $77°$
$L_3 = 40$ cm	$\theta = -90°$ to $90°$

9.2.5 INPUT DATA FOR THREE-DIMENSION PROBLEM

Length if links	Range of angles
$L_1 = 60$ cm	$\alpha = 20°$ to $360°$
$L_2 = 50$ cm	$\beta = -45°$ to $90°$
$L_3 = 40$ cm	$\theta = -90°$ to $90°$

9.2.6 IMPLEMENTATION OF GA

Binary Coding
To solve this problem using GA, a 14 digit binary coding is chosen to represent variables ($\alpha = 5$ digit, $\beta = 5$ digit and $\theta = 4$ digit),

11100 10001 0000

$$\text{Parameter value} = \text{Min value} + \text{DCV} \times \text{accuracy}$$

$$\text{Accuracy} = \text{Upper limit} - \text{lower limit}/2^n - 1$$

9.2.7 REPRODUCTION

- Chromosomes are selected from a population to be the parents to cross over and produce offspring.
- Various methods of selecting the chromosomes are present.
- The tournament selection method has been chosen.

9.2.8 TOURNAMENT SELECTION

- This method is one of the strategies to select the individuals from the population and insert them into the mating pool.
- Individuals from the mating pool are used to generate new offspring.

TABLE 9.2

Robot Coordinate System Using Genetic Algorithm (Two-Axis) — Input and Output Values of X, Y

S	Input X, Y (cm)	Output 1, X (cm)	Output 2, Y (cm)	Alpha, α (°)	Beta, β (°)	Theta, θ (°)
1	50, 30	49.9895	30.0138	86.0614	12.1658	30.9965
2	40, 40	39.9691	40.0297	110.8244	14.2341	45.0659
3	30, 50	30.0165	49.9904	99.3685	12.9743	59.047
4	20, 60	19.9552	59.9372	90.0003	19.2358	71.6218

- Individuals: 1   2   3   4   5   6
- Fitness: 1   2.10   3.11   4.01   4.66   1.91
- Parameter value = 0.75
- Step 1: Select two individuals, 2 and 3.
- No. 3 has a higher fitness.
- Generate a random number.
- If the random number is less than or equal to 0.75, select No. 3; else select No. 2.

9.2.9 Genetic Operators

Population size:100
Crossover probability (P_c): 0.7 (single point)
Mutation probability (P_m): 0.01
Number of generations: 100

Results obtained for the two-axis problem are given in Table 9.2. Results obtained for the three-axis problem are given in Table 9.3. GA Solution history is given in Figures 9.3 to 9.9.

TABLE 9.3

Robot Coordinate System Using Genetic Algorithm (Three-Axis) — Input and Output Values of X, Y

S	Input X, Y, Z (cm)	Output X (cm)	Output Y (cm)	Output Z (cm)	Theta, θ (°)	Alpha, α (°)	Beta, β (°)
1	60, 20, 30	59.9704	20.033	30.015	69.86	108.45	69.86
2	50, 30, 40	50.0501	30.044	40.017	72.852	120.99	21.39
3	40, 40, 50	40.0421	40.0421	49.952	68.544	135.02	32.93
4	30, 50, 60	29.9912	49.956	60.027	57.55	149.05	42.56

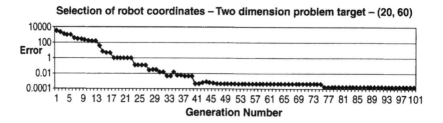

FIGURE 9.3 GA results – Selection of robot coordinates – Two dimension problem target – (20, 60).

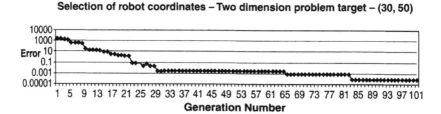

FIGURE 9.4 Selection of robot coordinates – Two dimension problem target – (30, 50).

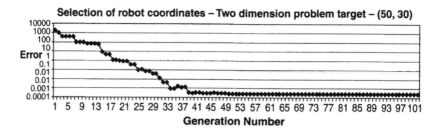

FIGURE 9.5 Selection of robot coordinates – Two dimension problem target – (50, 30).

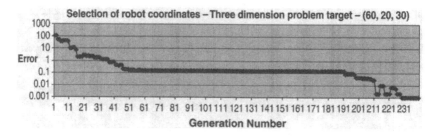

FIG.URE 9.6 Selection of robot coordinates – Three dimension problem target – (60, 20, 30).

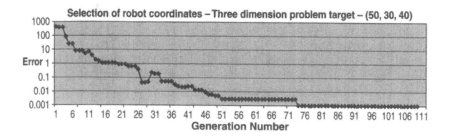

FIGURE 9.7 Selection of robot coordinates – Three dimension problem target – (50, 30, 40).

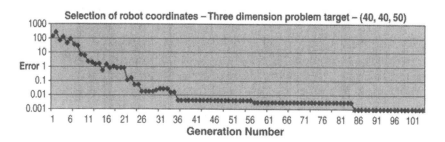

FIGURE 9.8 Selection of robot coordinates – Three dimension problem target – (40, 40, 50).

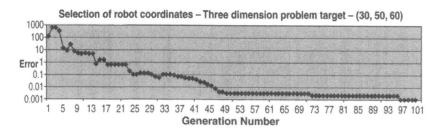

FIGURE 9.9 Selection of robot coordinates – Three dimension problem target – (30, 50, 60).

9.3 TRAJECTORY PLANNING FOR ROBOT MANIPULATORS USING GENETIC ALGORITHM

Trajectory is the plath traced in space by the end effector of the robot. It refers to the time history of position of the end effector between its start and end points. Ineg et al. presented a near optimal trajectory planning method for industrial robot manipulators.

In this simulation method, the reference trajectories are first developed using the robot dynamics equations and then the reference trajectories are evaluated based on the total energy consumed by the actuator. Another researcher proposed path planning by decomposition. Decomposition means the splitting of the robot into several chains, which in turn is a combination of several consecutive links and joints. A collision-free path can be obtained by refining the path obtained for each chain. Thus, decomposition minimizes the exponential growth of computation with robot degrees of freedom. This method also has certain limitations. It loses its utility for a low degree of freedom manipulator due to the overhead and it loses its effectiveness for a manipulator with many long links.

Xiangrong et al. presented a method for robot continuous path motion trajectories specified by a group of parameter equations in Cartesian coordinates. The time interval $(0, T)$ is divided into m segments and can be obtained in recurrence form. Recently, some researchers proposed GA as an optimization tool, using the minimum consumed energy as the criterion for trajectory generation. Hwang et al. addressed a global trajectory planning technique, which employs a collision trend index and a propagating interface model to perform mobile robot navigation. To simplify the mathematical representation and geometrical approximation, all the objects in the workspace are modeled as ellipses. The index is obtained by mapping the general relation between the ellipses into the profile of a Gaussian distribution. Jeyanthi et al. presented trajectory planning for a two-link planar manipulator using a random search method, which calculates the effect of joint angles on the time travel. Garg et al. used GA for torque minimization of robot path planning.

9.3.1 Problem Description

9.3.1.1 Robot Configuration

Two-link, three-link, four-link, five-link and six-link robot configurations are considered (Two-link robot is given in Figure 9.10). The velocity and acceleration of each link are taken as variables.

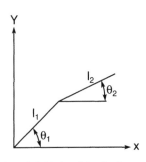

2-Link Robot Manipulator

FIGURE 9.10 Two-link robot manipulator.

Robot positions during motion are:

Two-link robot manipulator
  Initial position: $\theta_1 = 45°$, $\theta_2 = 45°$
  Final position: $\theta_1 = -60°$, $\theta_2 = 35°$

Three-link robot manipulator
  Initial position: $\theta_1 = 45°$, $\theta_2 = -60°$, $\theta_3 = 30°$
  Final position: $\theta_1 = 30°$, $\theta_2 = 35°$, $\theta_3 = 90°$

Four-link robot manipulator
  Initial position: $\theta_1 = 45°$, $\theta_2 = 60°$, $\theta_3 = 30°$, $\theta_4 = 90°$
  Final position: $\theta_1 = 30°$, $\theta_2 = 35°$, $\theta_3 = 90°$, $\theta_4 = 90°$

Five-link robot manipulator
  Initial position: $\theta_1 = 45°$, $\theta_2 = 60°$, $\theta_3 = 30°$, $\theta_4 = 90°$, $\theta_5 = 45°$
  Final position: $\theta_1 = 30°$, $\theta_2 = 35°$, $\theta_3 = 90°$, $\theta_4 = 90°$, $\theta_5 = 120°$

Six-link robot manipulator

  Initial position: $\theta_1 = 45°$, $\theta_2 = 60°$, $\theta_3 = 30°$, $\theta_4 = 90°$, $\theta_5 = 45°$, $\theta_6 = 80°$
  Final position: $\theta_1 = 30°$, $\theta_2 = 35°$, $\theta_3 = 90°$, $\theta_4 = 90°$, $\theta_5 = 120°$, $\theta_6 = 45°$

9.3.1.2 Estimation of Time

Given the following data as input for the problem, the procedure to determine the time for moving the 3-DOF end effector from initial position to final position is described below:

9.3.1.3 Input Data

Link lengths, l_i (l_1, l_2, l_3)
Initial joint angles, θ_i (θ_{1i}, θ_{2i}, θ_{3i})
Final joint angles, θ_f (θ_{1f}, θ_{2f}, θ_{3f})
Velocity characteristics of joints, v_i ($v_1 = d\theta_1/dt$, $v_2 = d\theta_2/dt$, $v_3 = d\theta_3/dt$)
Acceleration characteristics of joints, a_i ($a_1 = d^2\theta_1/dt^2$, $a_2 = d^2\theta_2/dt^2$, $a_3 = d^2\theta_3/dt^2$)

9.3.1.4 Procedure

Step 1: Choose a suitable value for v_1, v_2, and v_3.
Step 2: Choose a suitable value for a_1, a_2, and a_3.

Step 3: Determine the difference between the initial angle and final angle for each joint

$$\Delta\theta_1 = \theta_{1i}\ \theta_{1f}\ ;\ \Delta\theta_2 = \theta_{2i}\ \theta_{2f}\ ;\ \Delta\theta_3 = \theta_{3i}\ \theta_{3f}$$

Step 4: Determine the time required for providing the required rotation of each joint. This involves three time estimates:

- Time during acceleration (T_{ai})
- Time during constant velocity (T_{vi})
- Time during deceleration ($T_{di} = T_{ai}$)

$$T_i = T_{ai} + T_{vi} + T_{di}$$

$$T_i = (v_i/a_i) + [(\Delta\theta_i - 2\theta_{im})/v_i] + (v_i/a_i) \tag{9.1}$$

$$\theta_{im} = (v_i^2 - u_i^2)/2a_i$$

where
θ_{im} = Angle of rotation of joints during acceleration period
v_i = Maximum/final velocity
u_i = Initial velocity

Here, $\theta_{im} = v_i^2/2a_i$ $(\therefore u_i = 0)$. Two possibilities arise in the time calculation:

Change in rotation ($\Delta\theta_i$) is less than or equal to twice θ_{im} ($\Delta\theta_i \leq 2\theta_{im}$).
Change in rotation ($\Delta\theta_i$) is greater than twice θ_{im} ($\Delta\theta_i > 2\theta_{im}$).

Case 1: $\Delta\theta_i \leq 2\theta_{im}$. Figure 9.6 illustrates this situation. In such cases, the required change in rotation of $\Delta\theta_i$ is accomplished without constant velocity rotation, and so the second term of the time Equation 9.1 becomes zero.

$$T_i = T_{ai} + T_{di} = 2T_{ai} = 2(v_i/a_i) \tag{9.2}$$

Case 2: $\Delta\theta_i > 2\theta_{im}$. Figure 9.7 illustrates this situation. In such cases, the required change in rotation of $\Delta\theta_i$ is accomplished with constant velocity rotation.

$$T_i = T_{ai} + T_{ui} + T_{di} = 2T_{ai} + T_u i$$

$$T_i = 2(v_i/a_i) + [(\Delta\theta_i - 2\theta_{im})/v_i] \tag{9.3}$$

Calculate T_1, T_2, and T_3 from Equation 9.2 or Equation 9.3, as the case may be.
Step 5: Find the estimated time for the required movement.

$$T = \text{Max}(T_1, T_2, T_3) \tag{9.4}$$

The above steps are applied for different values of variables and estimate the minimum time among maximum (T_1, T_2, T_3) values. The variable values v_1, v_2, v_3, a_1, a_2, and a_3 corresponding to the minimum time are called optimal.

9.3.1.5 Assumptions

No restrictions on the payload of each joint.
The end effector moves without hitting any obstacles.
Joint motor characteristics are assumed.

Acceleration range for each joint is assumed.
Velocity range for each joint is assumed.
Acceleration time and deceleration time are equal.

9.3.1.6 Optimization Model

The time–optimal trajectory planning problem can be written as the following nonlinear optimization problem.

$$\text{Objective function} = \text{Min } t_f \qquad (9.5)$$

Subject to constraints

$$\theta(0) = \theta_0$$
$$v(0) = 0$$
$$\theta(1) = \theta_f$$
$$v(1) = 0 \qquad (9.6)$$

Variables are

$$v_{i,min} < v_i < v_{i,max} \quad \{i = 1, 2, 3...6\}$$
$$a_{i,min} < a_i < a_{i,max} \quad \{i = 1, 2, 3...6\}$$

Variable Ranges:

Two-link robot manipulator

$$30.0000 < v < 90.0000$$

$$30.0000 < a < 90.0000$$

Three-link, four-link, five-link robot manipulator

$$30.0000 < v < 90.0000$$

$$30.0000 < a < 180.0000$$

Six-link robot manipulator

$$30 < v < 90$$

$$30 < a < 120$$

9.3.1.7 Genetic Operators

Population size: 80
Total number of generations: 100
Reproduction: tournament selection
Crossover probability (P_c): 0.6000
Mutation probability (P_m): 0.0100

9.3.1.8 Simulation Results

Results for all configurations are given in Table 9.4 through Table 9.8. The genetic algorithm result histories are shown in Figure 9.11 through Figure 9.15.

TABLE 9.4
Two-Link Robot Manipulator

		Optimal Variable Values			
S	v_1 (°/sec)	a_1 (°/sec²)	v_2 (°/sec)	a_2 (°/sec²)	Optimum Time (sec)
1	64.36950	90.00	86.12904	90.00	2.055556

TABLE 9.5
Three-Link Robot Manipulator

			Optimal Variable Values				Optimum Time (sec)
S	v_1 (°/sec)	a_1 (°/sec²)	v_2 (°/sec)	a_2(°/sec²)	v_2 (°/sec)	a_3 (°/sec²)	
1	79.32552	160.6451	90.00000	180.00	85.54252	165.63049	1.555556

TABLE 9.6
Four-Link Robot Manipulator

				Optimal Variable Values					
S	v_1 (°/sec)	a_1 (°/sec²)	v_2 (°/sec)	a_2 (°/sec²)	v_3 (°/sec)	a_3 (°/sec²)	v_4 (°/sec)	a_4 (°/sec²)	Optimum Time (sec)
1	50.97707	74.1349	69.530	158.0058	63.7829	128.826	90.0000	180.000	2.50000

TABLE 9.7
Five-Link Robot Manipulator

| | Optimal Variable Values | | | | | | | | | | Optimum Time |
| | v_1 ($°$/sec) | a_1 ($°$/sec^2) | v_2 ($°$/sec) | a_2 ($°$/sec^2) | v_3 ($°$/sec) | a_3 ($°$/sec^2) | v_4 ($°$/sec) | a_4 ($°$/sec^2) | v_5 ($°$/sec) | a_5 ($°$/sec^2) | Time (sec) |
S											
1	44.31	149.6	82.31	128.6	63.37	159.6	90.00	180.0	58.56	70.76	2.500

TABLE 9.8
Six-Link Robot Manipulator

| | Optimal Variable Values | | | | | | | | | | | | |
S	v_1 ($°$/sec)	a_1 ($°$/sec^2)	v_2 ($°$/sec)	a_2 ($°$/sec^2)	v_3 ($°$/sec)	a_3 ($°$/sec^2)	v_4 ($°$/sec)	a_4 ($°$/sec^2)	v_5 ($°$/sec)	a_5 ($°$/sec^2)	v_6 ($°$/sec)	a_6 ($°$/sec^2)	Time (sec)
1	82.14	104.9	73.92	114.7	36.45	111.9	90.00	120.0	38.85	91.58	68.53	82.87	2.75

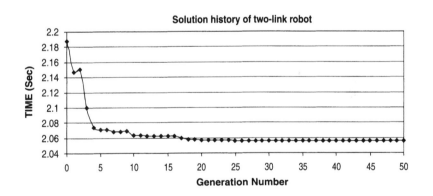

FIGURE 9.11 Solution history using GA – Two-link robot manipulator.

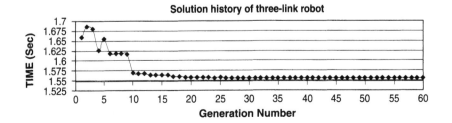

FIGURE 9.12 Solution history using GA – Three-link robot manipulator.

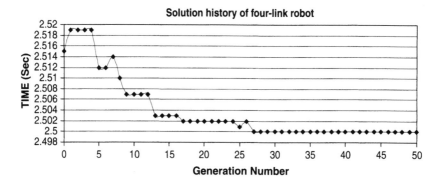

FIGURE 9.13 Solution history using GA – Four-link robot manipulator.

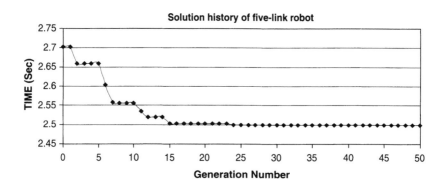

FIGURE 9.14 Solution history using GA – Five–link robot manipulator.

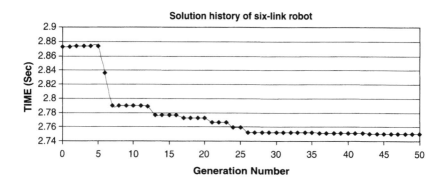

FIGURE 9.15 Solution history using GA – Six-link robot manipulator.

9.4 APPLICATION OF INTELLIGENT TECHNIQUES FOR ADAPTIVE CONTROL OPTIMIZATION

9.4.1 ADAPTIVE CONTROL SYSTEM (ACS)

An adaptive control system (ACS) can be defined as a feedback control system intelligent enough to adjust its characteristics in a changing environment in order to operate in an optimal manner according to some specified criteria. Generally speaking, adaptive control systems have achieved great success in aircraft, missile, and spacecraft control applications. Traditional adaptive control methods are suitable mainly for (1) mechanical systems that do not have significant time delays; and (2) systems designed so that their dynamics are well understood.

However, in industrial process control applications, traditional adaptive control has not been very successful. The most credible achievement is probably the above-described PID self-tuning scheme that is widely implemented in commercial products but not very well used or accepted by the user.

9.4.2 ADAPTIVE CONTROL OPTIMIZATION SYSTEM (ACOS)

ACOS uses some artificial intelligence systems like computers, sensors, signal processors, automation, and instrumentation systems to control the variables. The major problems with such systems have been difficulties in defining realistic indexes of performance and the lack of suitable sensors that can reliably measure the necessary parameters online in a production environment. The block diagram is shown in Figure 9.16.

CNC machine tools operate during the single working cycle with a constant, preprogrammed feed rate (f), although cutting conditions can vary during cutting operations (i.e., various cutting rates, milling wraps, or material structure). Adaptive control systems monitor the actual cutting conditions in real-time, modifying the feed rate automatically to the highest feasible value for maximum efficiency of each operation. An average savings of 10 to 15% in operation time has been achieved and for certain operations, and a savings of up to 30% in operation time has been achieved.

ACOS can yield increased tool life as well as tool breakage prevention and spindle overload protection. With the adaptive control systems, the cutting parameters can be adapted in real-time to the actual cutting conditions. During extreme overload conditions — for example, with contact between the tool and the workpiece (material entry) or a sudden, extreme increase in material hardness (e.g., hard inclusions in cast parts) or the cutting rate (e.g., over-measure variations) — the feed rate f is reduced automatically in real-time to the optimum value for these cutting conditions. When the cutting conditions return to normal, the feed rate is increased to the maximum feasible speed.

The control system identifies conditions in which the spindle load approaches the maximum permitted value (taking into account cutting tool and workpiece material characteristics) and stops the machine if necessary to prevent damage

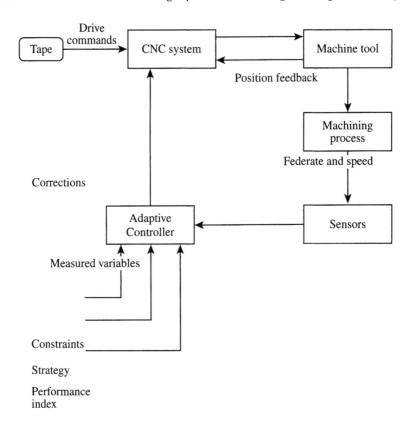

FIGURE 9.16 Adaptive control system for machine tool.

to the cutting tool, spindle, and the machine. At the same time, an alarm will be activated to notify the operator.

9.4.3 Application of Intelligent Techniques for ACOS

With this method, the important requirement is to find the suitable technique for the optimum operating parameters according to the condition of the machine tool such as cutting force, tool-chip interface temperature, power consumption, vibration, surface roughness of the product, tool wear, and so on.

Five techniques are described in Chapter 3. All these techniques can be applied for the above application. Optimization procedures using genetic algorithm and tabu search have been described in Chapter 6 for finding the operating parameters for CNC machine tools. These procedures can be suitably modified for the use of ACOS.

REFERENCES

Boctor, F.F., A linear formulation of the machine-part cell formation problem, *International Journal of Production Research,* 29(2), 343–356, 1991.

Capi, G., Kaeko, S., Mitobe, K., Barolli, L., and Nasu, Y., Optimal trajectory generation for a biped robot using genetic algorithm, *Journal of Robotics and Autonomous Systems,* 38(2), 119–128, 2002.

Garg, D.P. and Kumar, M., Optimization techniques applied to multiple manipulators for path planning and torque minimization, *Engineering Applications of Artificial Intelligence,* 15, 241–252, 2002.

Groover, M.P., *Industrial Robotics,* McGraw-Hill, Singapore, 1987.

Hartly, J. *Robots at Work,* IFS Publishing, HAR, U.K., 1983.

Hourtash, A. and Tarokh, M. Manipulator path planning by decomposition algorithm and analysis, *IEEE Transactions on Robotics and Automation,* 17(6), December 2001.

Hwang, K.-S. and Ming-Yi-Ju, A propagating interface model strategy global trajectory planning among moving obstacles, *IEEE Transactions on Industrial Electronics,* 49(6), December 2002.

Ineg, D.Y. and Chen, M., Robot trajectory planning using simulation, *Journal of Robotics and Computer Integrated Manufacturing,* 3(2), 121–129, 1997.

Jeyanthi, G.L. and Jawahar, N., Minimum time trajectory for 2-dimensional 3-degree of freedom (2D-3DOF) robot, *Proceedings of the National Conference on Modeling and Analysis of Production Systems,* Department of Production Engineering, NIT, Tiruchirappalli, India, 149–155, 2003.

Katz, Z. and van Niekerk, T., Implementation aspects of intelligent machining, *Proceedings of the Institution of Mechanical Engineers,* 217, 601–613, 2003.

Moshan, S., *A Robot Engineering Textbook,* Harper & Row, New York, 1987.

Xiangrong, Xu. and Chen, Y., A method for trajectory planning of robot manipulators in Cartesian space, *IEEE Proceedings on Intelligent Control and Automation,* Hefei, China, July 2, 2000.

10 Conclusions and Future Scope

Due to tough global market competition, manufacturers must produce products effectively to receive high customer satisfaction. To achieve this, they must produce a variety of products to meet the expectations of different segments of the market. The investment cost is high to update existing production facilities and other development activities in addition to a high operating cost. On the other hand, manufacturers need to meet certain objectives like minimizing the product cost, maximizing the product quality, or a combination of both or other objectives to compete with other products in the global market. The current requirements are to utilize resources, such as manpower, raw materials, and equipment, effectively and efficiently, to improve their performance. A need exists to optimize all the manufacturing functions.

To meet the above requirements, we must develop and implement the new area called manufacturing optimization. The main purpose of this book is to create an awareness about this subject among engineering and management students, research scholars, faculty, and practicing engineers and managers. To implement and realize the benefits from this subject a strong industry–academia interaction is required. Academic people are good in the application of new techniques and the development of new procedures and systems. Practicing engineers and managers are good in the understanding of real-world applications and problems. With the joint effort of the academic community and practicing engineers, better methods and procedures can be developed for real-world manufacturing applications.

The majority of manufacturing optimization problems are multivariable, nonlinear, constrained problems. Several conventional optimization techniques are available for solving different manufacturing optimization problems but they are not robust and several have difficulty in implementing these techniques. If a particular conventional technique is good at solving a particular type of problem, we can apply that technique to get the optimal solution. Otherwise, if problems are encountered with conventional techniques or it is not possible to apply them, the intelligent techniques described in this book can be pursued.

Five new techniques are described in this book with several examples taken from the literature for different types of manufacturing optimization problems: design, tolerance allocation, selection of machining parameters, integrated product development, scheduling, concurrent formation of machine groups and part families, selection of robot coordinates, robot trajectory planning, and intelligent machining. All the manufacturing functions described in this book have been successfully solved by genetic algorithm techniques. Other intelligent techniques have been implemented only for solving certain types of problems: simulated annealing; design and scheduling, particle swarm optimization and ant colony optimization; tolerance allocation, tabu search; machining parameters optimization.

Comparative analysis is given for conventional and intelligent techniques for certain problems. But the analysis is not given for the selection of a particular intelligent technique from the five techniques described in this book. It is left to the reader to select a suitable intelligent technique for solving the various manufacturing optimization problems that they encounter. The genetic algorithm can be applied for solving any type of manufacturing optimization problems. Other techniques have not been tried for all the problems described in this book; only nine manufacturing activities are included in this book. A lot of scope remains for research in this area and the application of these techniques and procedures to other manufacturing activities, like maintenance, quality control, plant layout design, line balancing, automated monitoring, process planning, production planning control, and many other topics.

Index

Printed and bound by CPI Group (UK) Ltd, Croydon, CR0 4YY

18/10/2024

01776242-0001